Georg Klemperer

The Elements of Clinical Diagnosis

Georg Klemperer

The Elements of Clinical Diagnosis

ISBN/EAN: 9783743362031

Manufactured in Europe, USA, Canada, Australia, Japa

Cover: Foto ©berggeist007 / pixelio.de

Manufactured and distributed by brebook publishing software (www.brebook.com)

Georg Klemperer

The Elements of Clinical Diagnosis

ELEMENTS OF CLINICAL DIAGNOSIS

THE ELEMENTS

OF

CLINICAL DIAGNOSIS

BY

PROFESSOR DR. G. KLEMPERER

PROFESSOR OF MEDICINE AT THE UNIVERSITY OF BERLIN

First American from the Seventh (last) German Edition

WITH SIXTY-ONE ILLUSTRATIONS

AUTHORIZED TRANSLATION

BY

NATHAN E. BRILL, A.M., M.D.

ADJUNCT ATTENDING PHYSICIAN, MOUNT SINAI HOSPITAL, NEW YORK CITY

AND

SAMUEL M. BRICKNER, A.M., M.D.

ASSISTANT GYNÆCOLOGIST, MOUNT SINAI HOSPITAL, OUT-PATIENT DEPARTMENT

New York

THE MACMILLAN COMPANY

LONDON: MACMILLAN & CO., LTD.

1898

All rights reserved

DEDICATED TO

Privy Counsellor Professor Dr. E. Leyden

DIRECTOR OF THE FIRST MEDICAL CLINIC

ON THE

OCCASION OF THE JUBILEE IN CELEBRATION OF

THE COMPLETION OF HIS

TWENTY-FIVE YEARS OF SERVICE

AS A CLINICAL TEACHER

APRIL 6, 1890

AUTHOR'S PREFACE TO THE FIRST AMERICAN EDITION

Of the great number of American physicians who annually visit the clinics of Germany, not a few have taken my diagnostic course, the fundamental principles of which are contained in this little book.

May this translation renew the memories of mutual work in the minds of many of my colleagues, and add its modest share in strengthening the relations between American and German intellectual life.

G. KLEMPERER.

BERLIN, November 30, 1897.

AUTHOR'S PREFACE TO THE FIRST EDITION

In this little book the rules governing the clinical diagnosis of medical cases which are in vogue in the first medical clinic of Berlin, are repeated. The same methods have characterized my didactic work since its inception.

I am but returning that which I learned in my happy student days, when I add this modest little book to the offerings of homage which my esteemed chief will receive from his thankful pupils on the day of his clinical jubilee.

G. KLEMPERER.

BERLIN, March 15, 1890.

AUTHOR'S PREFACE TO THE SEVENTH EDITION

The present edition has undergone several additions and changes which were made necessary by the progress in the diagnosis of internal diseases.

The chapter on the Röntgen rays is new.

G. KLEMPERER.

BERLIN, July 15, 1897.

TRANSLATORS' PREFACE

In offering Klemperer's well-known work to the American medical public, we feel that no better opportunity than this could be found to express our own admiration for the excellence of the book in the original language. The fact that in seven years the book has seen as many editions in German, certainly calls for no further comment upon its virtue from us.

We have followed the original plan of Klemperer's work throughout, retaining not only its form, but most of the illustrations and the continental terms. We feel justified in adhering to the Latin names of diseases and of technical terms, since it will familiarize students with them, and facilitate their subsequent medical reading and study. The simple explanation that we desired to maintain, as far as possible, the original form of the work, is sufficient apology —if apology need be— for using the terse phraseology in descriptions of diseases and conditions.

Where new illustrations have been substituted, it has been done only for the sake of clearness.

If the English translation of Klemperer's work shall prove as helpful to the medical profession of the United States as its German analogue has to us, we shall feel amply repaid for our labor.

N. E. BRILL,
S. M. BRICKNER.

NEW YORK, December 1, 1897.

TABLE OF CONTENTS

	PAGE
THE METHOD OF DIAGNOSTIC EXAMINATION	1

CHAPTER I

ANAMNESIS AND GENERAL CONDITION	4

CHAPTER II

Diagnosis of the Acute Febrile and Acute Infectious Diseases	16
Special Symptomatology	21

CHAPTER III

Diagnosis of the Diseases of the Nervous System	33
Special Symptomatology	67

CHAPTER IV

Diagnosis of Diseases of the Digestive System	70
Diagnosis of Stomach Diseases	75
Special Symptomatology	90
Diagnosis of the Diseases of the Intestines and Peritoneum	91
Special Symptomatology	99
Diagnosis of the Diseases of the Liver	100
Special Symptomatology	103
Diagnosis of Enlargement of the Spleen	106

CHAPTER V

	PAGE
DIAGNOSIS OF THE DISEASES OF THE UPPER AIR-PASSAGES (NOSE, THROAT, LARYNX)	107
Special Symptomatology	114

CHAPTER VI

DIAGNOSIS OF THE DISEASES OF THE RESPIRATORY TRACT	121
Percussion of the Thorax	129
Auscultation of the Thorax	136
Examination of the Sputum	141
Special Symptomatology	151

CHAPTER VII

DIAGNOSIS OF DISEASES OF THE CIRCULATORY SYSTEM	160
The Pulse	171
Special Symptomatology	176

CHAPTER VIII

EXAMINATION OF THE URINE	181

CHAPTER IX

DIAGNOSIS OF DISEASES OF THE KIDNEY	220
Diffusive Diseases of the Kidneys	220
Other Diseases of the Kidneys	224
Examination of Renal and Vesical Calculi	226

CHAPTER X

DIAGNOSIS OF DISORDERS OF METABOLISM	229

CHAPTER XI

	PAGE
DIAGNOSIS OF DISEASES OF THE BLOOD . .	241
Special Symptomatology	250

CHAPTER XII

ANIMAL AND VEGETABLE PARASITES 253

CHAPTER XIII

THE RÖNTGEN RAYS AS DIAGNOSTIC AIDS 278

The "Status of the Nervous System," on page 33, is edited by Professor Dr. GOLDSCHEIDER of Berlin.

The illustrations were drawn by Dr. JOHANNES MANN of Leopoldshall, mostly from his own specimens, in part from older specimens and drawings of Professor LEYDEN.

Figures 53 to 61 have been taken, by permission, from the microphotographic Atlas of FRÄNKEL and PFEIFFER.

LIST OF ILLUSTRATIONS

FIG.		PAGE
1.	Fever Curve in Measles	21
2.	Fever Curve in Scarlatina	22
3.	Fever Curve in Erysipelas	22
4.	Fever Curve in Pneumonia; crisis 4th to 5th day	23
5.	Fever Curve in Pneumonia; pseudocrisis on the 3d, crisis on the 11th day	23
6.	Schematic Fever Curve in Typhoid Fever	24
7.	Fever Curve in Typhus Fever	25
8.	Fever Curve in Relapsing Fever	26
9.	Fever Curve in Variola	26
10.	Fever Curve in Varioloid	27
11.	Fever Curve in Quotidian Intermittent Fever	28
12.	Fever Curve in Tertian Intermittent Fever	28
13.	Fever Curve in Quartan Intermittent Fever	29
14.	External Surface of the Left Cerebral Hemisphere	38
15.	Outline of a Transverse Dorso-ventral Section of the Right Half of the Brain	39
16.	Diagrammatic Cross-section of the Spinal Cord, showing the Tracts of the White Matter	41
17.	The Base of the Brain and the Cranial Nerves	46
18.	Points for Electric Stimulation of the Nerves of the Head, Face, and Neck	61
19.	Points for Electric Stimulation of the Nerves of the Flexor Surface of Arm, Forearm, and Hand	62
20.	Points for Electric Stimulation of the Nerves of the Extensor Surface of Arm, Forearm, and Hand	63

ILLUSTRATIONS

FIG.		PAGE
21.	Points for Electric Stimulation of the Nerves of the Extensor Surface of the Thigh and Leg	64
22.	Points for Electric Stimulation of the Nerves of the Flexor Surface of the Thigh, Leg, and Foot	65
23.	Schematic Microscopic Picture of Vomitus	80
24.	Relative Positions of the Stomach, Liver, and Colon	83
25.	The Various Paralyses of the Vocal Cords	117
26.	Relative Positions of Thoracic Viscera	130
27.	Morning Sputum of Chronic Bronchitis, containing no Pathological Elements	146
28.	Asthma Crystals	148
29.	Curschmann's Spirals	149
30.	Sphygmographic Tracing of the Radial Artery of a Healthy Young Man	175
31.	Sphygmographic Tracing of the Radial Artery in Aortic Insufficiency	175
32.	Sphygmographic Tracing of the Radial Artery in Aortic Stenosis	175
33.	Sphygmographic Tracing of the Radial Artery in Mitral Stenosis	175
34.	Improvised Fermentation Apparatus (Moritz)	195
35.	Urinary Deposits in Acid Urine	211
36.	Sediment of Urine in Acute Yellow Atrophy of the Liver	212
37.	Sediment of Ammoniacal Urine	213
38.	Sediment in Acute Nephritis	215
39.	Sediment in Chronic Nephritis	215
40.	The Blood in Pernicious Anæmia	243
41.	Schematic Representation of Various Kinds of Leucocytes	247
42.	Microscopic Picture of Tænia solium (Head, Proglottides, Egg)	255
43.	Microscopic Picture of Tænia saginata (Head, Proglottides, Egg)	256
44.	Microscopic Picture of Bothriocephalus latus (Head, Proglottides, Egg)	256

ILLUSTRATIONS

FIG.		PAGE
45.	Echinococcus Membrane and Hooklets	258
46.	Egg of Distoma hepaticum	258
47.	Egg of Distoma hæmatobium	258
48.	Egg of Ascaris lumbricoides	259
49.	Egg of Oxyuris vermicularis	259
50.	Egg of Anchylostoma duodenale	260
51.	Egg of Trichocephalus dispar	260
52.	Trichinæ in Muscle	261
53.	Staphylococci	268
54.	Streptococci	269
55.	Gonococci	269
56.	Pneumococci	271
57.	Pure Culture of Typhoid Bacilli	271
58.	Pure Culture of Cholera Bacilli	273
59.	Bacilli in Tubercular Sputum	274
60.	Spirilla from a Case of Relapsing Fever during the Fever	275
61.	Diphtheria Bacilli	276

THE ELEMENTS OF CLINICAL DIAGNOSIS

THE METHOD OF DIAGNOSTIC EXAMINATION

The tasks of practical medicine are prophylactic and curative. A systematic treatment of disease rests upon the recognition of its forms and manifestations. The science involved in the recognition of disease is called *diagnosis*.

A complete diagnosis includes: (1) The naming of the disease, *i.e.*, its classification or place in some particular group of diseases; (2) the recognition of the stage of the disease, peculiarities or complications; (3) the appreciation and estimation of dangers existing or liable to arise.

A diagnosis is reached by the examination of the patient. This consists in the obtaining of the history (anamnesis) and in the objective examination (status praesens). It is always advisable to follow the order of a systematic scheme in order to avoid mistakes of omission.

The following scheme of examination has long been used in the first medical clinic of Berlin:—

Name, age, civil condition. Date of examination.
Anamnesis:
1. Hereditary relations.
2. Diseases of childhood, menstruation.
3. General conditions of life, occupation.
4. Previous diseases, puerperia.
5. Present illness, its prodromata and suspected cause.

6. The first phenomena of the disease. (Chills and fever? Subjective complaints, functional disturbances.)
7. Course of the disease to date.
8. How long did it increase in severity? Has it been better or worse?
9. Previous treatment.
10. Complications: data of patient as to the principal functions; *e.g.*, sleep, appetite, cough, expectoration, urine, etc., condition of nutrition and strength, appearance.

Status præsens.
  A. *General Examination.*
    I. *Constitution* (stature, osseous system, muscles, fat).
    II. *Position* (active or passive, recumbent position, etc.).
    III. *Face.*
      1. Color (cheeks, lips, conjunctivæ).
      2. Nutrition, congestion.
      3. Expression.
      4. General appearance.
    IV. *Skin.*
      1. Color (p. 7).
      2. Eruptions, œdema, scars, decubitus.
      3. Moist or dry.
      4. Temperature and its distribution.
    V. *Pulse.*
      1. Frequency, rhythm.
      2. Condition of the arteries (straight, tortuous, sclerosed).
      3. Tension of the arteries.
      4. Pulse-wave.
    VI. *Respiratory frequency and rhythm.*
    VII. *Striking symptoms.*
    VIII. *Complaints of the patient.*
  B. *Special Examination.*
    I. *Nervous system.*
      1. Sensorium (free or stuporous).
      2. Headache, vertigo.
      3. Sleep.
      4. Tremor.
      5. Delirium, abnormal moodiness.
      6. Disturbances of sensation and motility.
    II. *Digestive system.*
      1. Lips, tongue.

2. Throat.
3. Appetite.
4. Thirst.
5. Vomiting.
6. Stools.
7. Palpation of the abdomen (painful areas? tumors?), the liver and spleen.
8. Percussion of the abdomen (stomach, liver, spleen, tumors).
9. Examination of stomach contents.

III. *Respiratory system.*
1. Rhythm of respiration.
2. Shape of thorax.
3. Respiratory movements (frequency, type, amplitude, unilateral).
4. Cough and expectoration.
5. Percussion.
6. Auscultation.
7. Pectoral fremitus and bronchophony.

IV. *Circulatory apparatus.*
1. Inspection of the cardiac area.
2. Inspection of the large vessels.
3. Palpation of the cardiac impulse.
4. Palpation of the apex beat.
5. Percussion of the heart.
6. Auscultation of the heart.
7. Auscultation of the large vessels.

V. *Urine.*
1. Voluntary, painful evacuation.
2. Amount in 24 hours.
3. Specific gravity.
4. Color, clearness.
5. Reaction.
6. Albumin, sugar.
7. Sediment and its contents.

Note.—The beginner will do well to learn this or some similar scheme well and to adhere to its order in all his examinations. The experienced physician determines the general condition in a few minutes while asking the patient the necessary questions. Through the general condition and the history, attention will be directed diagnostically toward certain organs or systems of organs with which the special examination is then begun. The organ found diseased must be examined with the greatest care: so far as the other organs are concerned, the determination of their main features is sufficient.

CHAPTER I

ANAMNESIS AND GENERAL CONDITION

1. Anamnesis. — An exact history of a case is of the greatest importance; for, frequently, the decision of a diagnosis hinges upon it.

Hereditary taint is especially significant in the diagnosis of phthisis and the nervous diseases.

The causes of a present illness may lie in the sequelæ of some previous disease; *e.g.*, scarlatina may be productive of an acute, more rarely of a chronic, nephritis; an articular rheumatism may lead to a valvular endocarditis; repeated attacks of bronchitis and asthma may superinduce emphysema.

Certain occupations may entail certain diseases: painters are subject to lead poisoning, cornetists to emphysema; compositors, millers, and stonecutters are liable to acquire phthisis.

Certain noxious substances possess etiological character: alcohol induces cirrhosis of the liver, or a weak heart, or multiple neuritis, or chronic nephritis.

Certain events in the history of a patient are of diagnostic value: *e.g.*, an hæmoptysis (phthisis), hæmatemesis (gastric ulcer), attacks of jaundice (gall-stones).

It is well for the beginner to remember that the taking of the history is often the occasion of the first meeting between patient and physician. To win the confidence of the patient, the questions should always be asked in a friendly and gentle manner, and modified to suit the patient and the occasion.

2. Condition of nutrition and strength. — The state of a patient's nutrition is usually easily recognized by a glance at the face (fat or thin, of pale or ruddy color, bright or sunken eyes, lively or depressed expression) and the rest of body (fat, muscular development of the buttocks, of the arms and legs).

The state of the nutrition leads the diagnosis to a particular group of diseases. A poorly nourished condition is indicative of the so-called *cachectic* diseases (phthisis, carcinoma, leucæmia, anæmia, the severer form of diabetes). A patient who maintains a well-nourished condition in spite of long-continued illness is not a victim of cachectic disease. Because of their short duration, the acute febrile diseases do not essentially disturb the nutrition; the sub-acute diseases (typhoid fever, meningitis), on the other hand, lead to marked emaciation.

The nutritive state is of particular importance in the differential diagnosis of the diseases of the lungs and stomach. Phthisis and carcinoma of the stomach are suspected in cachectic individuals; bronchitis, ulcer, and neurotic disturbances of the stomach appear more often in healthy-looking persons.

3. **Constitution and habitus.** — Through the repeated observation of the sick, the physician secures certain impressions of single forms of disease, by means of which, on seeing similar cases, he at once suspects the nature of the illness. These impressions are a composite of the nutrition, complexion, appearance, position, speech, etc. The judgment as to the habitus is of unquestionable worth, but must not preclude a careful examination.

Habitus phthisicus is seen in tubercular subjects. It is characterized by a pale, often spiritual, countenance, with a fine skin and circumscribed redness of cheeks; the neck is slender, the thorax weakly acting. The figure is thin and wasted, the hands are small and white.

Habitus apoplecticus is marked by a dark-red, round, stout face, glistening watery eyes, and short neck. The chest is usually barrel-shaped, and the respiration is of a short, wheezy character, indicative of emphysema. The body is abnormally fat. If the patient is alcoholic, he has a tendency to apoplectic insults.

With the *habitus neurasthenicus*, the patient is usually well nourished and has an expressive face. The eyes are bright, have an unsteady gaze, and are passive as though their possessor were in suffering. These patients are moody and hypochondriacal, and are often suspicious. Speech often impulsive.

To recognize the habitus associated with a particular disease is largely dependent upon experience, though it is sometimes an intuitive faculty.

4. **The position of the patient** in bed, which is immediately observed, may modify the general impression as to the character of the disease. If the patient is lying on his back, it must be observed whether he lies with little muscular effort (*active position*), or, yielding to his burdens, he is sunk down in bed, with knees drawn up (*passive position*). The latter position, if maintained, is always a sign of weakness or collapse, and is usually of grave significance.

A patient with disease on one side of the body (*e.g.*, pneumonia, pleurisy, pneumo-thorax) often assumes a *side* position, usually lying on the diseased side. In some gastric affections, the patient lies on the abdomen.

Restlessness or jactitation is a sign of fever (see below), and is often a forerunner of delirium. At the same time, the danger of such conditions as wakefulness and stupor must be regarded.

The *sitting posture* is most frequently observed in the disturbed compensation of cardiac disease, and is a symptom of dyspnœa or orthopnœa.

5. **The features, expression, and appearance** of the patient give valuable hints in the determination of a diagnosis.

The *facies composita* is the intelligent play of features seen in health. The *facies Hippocratica* or *decomposita* is the fixed, distorted, inanimate appearance of the features in unconsciousness and approaching death.

The physician should learn to estimate a patient's sensorium from his facial expression. In the febrile diseases this is of especial value and importance. Typhoid fever, meningitis, and acute miliary tuberculosis, sepsis, for instance, run their courses with disturbed mental and nervous functions, and the face is then expressionless, dull, and apathetic. The facial expression of patients suffering from such diseases is very characteristic and may be recognized at the first glance. In other febrile diseases, the face has a turgescent but a simultaneously clear and expressive appearance.

The beginner should learn thoroughly the various types of expression; the study of the physiognomy is of unquestionable diagnostic value, and was highly regarded and constantly practised by the physicians of bygone years, masters of observation. A careful examination must not, of course, be precluded.

6. **The complexion and the skin** may offer some help in establishing a diagnosis.

The normal pale-red color of the skin, with rosy cheeks and bright-red lips, allows only negative conclusions.

a. Flushing of the face, accompanied by heat, turgescence, sweat, and bright, bulging eyes, is usually a sign of fever. Further symptoms of temperature would, of course, be looked for; but one must not forget the fleeting redness due to excitement and modesty.

b. Abnormal *paleness* (*pallor eximius*), chalky or waxy color of the cheeks and body, pallor of the lips and conjunctivæ is a sign of anæmia. The anæmia may be *primary* (essential anæmia) or *secondary*. It is primary when dependent upon disease of the blood, as proven by an examination of the blood (Chap. XI.). It is secondary when it follows or is coincident upon severe systemic disease (phthisis, carcinoma, amyloid degeneration, etc.). The sudden appearance of an anæmia which has a continued duration and which is preceded by collapse, speaks in favor of an internal hæmorrhage: in the stomach, in the intestine, or in the Fallopian tube.

c. Jaundice (*icterus*) is a sign of the deposit of bile-pigment in the skin and is usually coincident with some form of *liver* disease. A jaundice occurring in a well-nourished patient usually denotes a catarrhal inflammation of the duodenum and bile-ducts (*icterus simplex*). Jaundice with severe illness is indicative, as a rule, of grave affections of the liver (*icterus gravis*, Chap. IV.). Independent of liver diseases, the graver form of jaundice may result from toxic absorption, which destroys the red blood cells (*cythæmolysis*), as in acute phosphorus poisoning, or from the products of the acute infectious diseases.

The yellow color of the skin may be induced by the administration of *picric acid;* but then the urine will not respond to the tests for bile-pigments.

The classification into *icterus simplex* and *icterus gravis* is of great practical value. Of more scientific importance, however, is the division into jaundice with and without *polycholia*. Icterus with polycholia is produced by an overflow of bile in consequence of great disturbance of the red blood corpuscles. It is comparatively rare, and leads rapidly either to death or to recovery and is never chronic. The stool is always well colored with this form of jaundice. Jaundice without polycholia is caused by a congestion in the bile-ducts and occasions the simple jaundice and the greater number of cases of malignant jaundice. In these cases the stools are usually colorless, the color varying with the degree of occlusion of the bile-ducts.

d. A *bronze-colored* skin is the pathognomonic symptom of Addison's disease, which has its probable lesions in the suprarenal capsules and splanchnic nerves. It is always fatal, and is accompanied by an increasing cachexia. The brown pigmentation of the mucous membrane of the mouth is particularly characteristic.

e. Cyanosis is the reddish-blue condition of the tissues seen in certain diseases, most easily recognized in the lips and finger-nails. It depends upon an over-charge of carbon dioxide in the blood. It may be caused by: 1. Slow or imperfect circulation of the blood, and it is, therefore, an important symptom of uncompensated *cardiac* disease. 2. Disturbance in the exchange of gases in the lungs by (*a*) pulmonary disease, or (*b*) abnormal distention of the abdomen by tumors, meteorism, or ascites. These conditions lead to cyanosis, as a rule, only in the advanced stages, whereas at the beginning, or at the height of the disease, there is frequently a vicarious increase in the force of respiration in the unaffected portion of lung. In *pneumonia* the appearance of cyanosis is of grave import. In *acute miliary tuberculosis* there is usually very pronounced cyanosis.

A local cyanosis arises from venous congestion by thrombi or tumors. In the face, it may be the result of freezing.

A combination of cyanosis and pallor (*livor*) is sometimes seen in patients exhausted by heart disease. Organic cardiac disease, with deep congestion, not uncommonly produces cyanosis with jaundice, the latter arising from congestion of, and absorption from, the gall-ducts in the congested liver.

7. **Dyspnœa** (air-hunger, impeded respiration). — The physician must determine at once whether the respiration is normal or not, but should notice especially whether it is free or labored.

It is essential to differentiate between *increased* and *labored* respirations. A simple increase in the number of respirations, above 24 to the minute, is of no particular diagnostic value, and can appear after bodily exertion, as a result of emotional influences, in hysteria or in fever.

Dyspnœa is an indication of the need of oxygen. It is characterized by rapidity and deepening of the respiratory movements, with contractions of the auxiliary muscles of respiration. The patient feels "air-hungry" up to the point of suffocation. True dyspnœa is usually accompanied by cyanosis, a condition pathognomonic of uncompensated heart disease and advanced disease of the lungs. Less commonly they appear together in some abdominal diseases which impede diapragmatic movement.

Attacks of dyspnœa which pass away more or less rapidly, and are followed by longer or shorter periods of quiet breathing, are called asthmatic attacks.

In *Cheyne-Stokes* respiration, the respiratory movements gradually decrease both in extent and rapidity until they cease altogether, and a condition of apnœa ensues. This is followed by a feeble respiration, succeeded in turn by a somewhat stronger one. The amplitude of the respiratory movements increases and wanes again in a similar manner. This phenomenon is seen especially in uræmia, also in cardiac and brain diseases, and is usually of grave significance. Still, an indication of the Cheyne-Stokes phenomenon is sometimes seen in the sleep of healthy persons.

The description of other changes in respiration is reserved for the chapter devoted to the subject (p. 121).

8. The excessive accumulation of lymph in the subcutaneous connective tissues is known as **œdema** or **dropsy**. The skin above a dropsical part of the body remains pitted after pressure by the finger. Œdema is such a striking symptom that patients frequently make it their principal complaint; stupid patients may overlook it, however, and it is a good practice to accustom oneself to look for it.

The first signs of œdema can be recognized by pressure of the fingers in the neighborhood of the ankles.

The presence or absence of œdema determines the path of the diagnosis. If there are no complicating cyanosis and dyspnœa pointing to heart disease, the nutritive condition and the condition of the blood must be considered and the urine must be examined for albumin.

a. Dropsy with cyanosis and dyspnœa is the symptom of uncompensated heart disease: *cardiac* dropsy.

The diseases of the heart interfere in a marked degree with the venous return of the blood. The blood remains too long in the tissues, where it loses its O and takes up more CO_2 than ordinary. The over-distended veins no longer take up the lymph in the usual quantity, and the latter fluid floods the tissues.

b. Dropsy with albuminuria is the symptom of disturbed function of the kidneys, and is known as *renal* dropsy. Richard Bright (1825), an English physician, first described this combination of symptoms, and it is therefore called "Bright's disease."

The dropsy in albuminuria may be explained as follows: Normally, the capillaries are impermeable to large quantities of plasma on account of the vital activity of the cells in their walls. This activity is intact only when the cells are well nourished, *i.e.*, when the composition of the blood is normal. The blood becomes deteriorated when the kidneys are diseased; in health, these organs separate all waste products from the blood; in disease of the renal epithelium, metabolic products remain in the blood, the walls of the capillaries become permeable because of poor nutrition, and œdema results. At the same time, all diseases of the renal epithelium lead to the appearance of albumin in the urine.

Another theory attributes renal œdema mainly to an inflammatory process evoked by the action of the same noxious materials upon the walls of the capillaries as induces a similar process in the kidneys.

In œdema due to *great congestion*, there is often albuminuria because of the simultaneous congestion of the renal veins. The diagnosis is turned toward the heart by the accompanying cyanosis and dyspnœa.

In chronic diffuse nephritis with or without exudation and with the atrophic kidney with granular degeneration there is frequently a hypertrophy of the left ventricle of the heart. The heart and the kidney both give symptoms.

c. Œdema accompanying *cachexia* can be observed in all conditions of bad nutrition, especially in the cachectic diseases: carcinoma, pernicious anæmia, phthisis, diabetes. It can appear temporarily after excessive exertion and in inanition.

Under these conditions, the œdema can also be explained by the abnormal condition of the blood, the consequent disturbance of the intima, and the permeability of its cells. The impoverishment of the blood arises from disease of the blood (anæmia, severe chlorosis, leucæmia), or from ill-nutrition or from chronic organic disease. The transitory œdema of young people who have excessively exercised may be partly explained by their prolonged upright position.

9. **Exanthemata** (eruptions). — The examination of the skin is incomplete unless eruptions are searched for. These are of particular importance in the febrile diseases in which

they sometimes determine the diagnosis. It is difficult to recognize the various eruptions from description alone. The student should therefore improve every opportunity to observe them.

The eruption of *measles* is scattered and is composed of large macules; that of *scarlatina*, small macules, lying closely together so that it gives an appearance of diffuse redness. In *typhoid fever* there is a *roseola*, sparse over the abdomen, less thickly scattered over the chest, which may vary in size from the head of a pin to a pea. In *typhus*, a profuse roseola. Exanthemata may appear sometimes only several days after the beginning of the fever; the absence of the eruption does not, therefore, exclude the diagnosis of an exanthematous disease.

Exanthemata *without fever*, with little or no general disturbance, may be referred to diseases of the skin often dependent on syphilis.

Occasionally eruptions follow the exhibition of certain *drugs*, such as the measles-like eruption of antipyrin and the acne of potassium iodide. These disappear, naturally, after the cessation of the medication.

Subcutaneous hæmorrhages (ecchymoses, suggillations, or petechiæ when punctate) occur at times: (1) in severe forms of some of the acute infectious diseases, as scarlatina, small-pox, typhoid, typhus, hæmorrhagic measles; (2) in acute articular rheumatism, even in cases not particularly severe, peliosis, and purpura rheumatica; (3) in very severe anæmia or leucæmia; (4) in some diseases of the liver, as acute yellow atrophy, rarely in cirrhosis; (5) in certain diseases due to disturbed metabolism, which lead to great weakness: morbus maculosus Werlhoffi, scurvy. In the last named disease, bleeding from the gums is characteristic.

Petechiæ, which become pustular, are symptomatic of embolism in the skin in severe pyæmia, ulcerative endocarditis, and glanders.

10. **Temperature of the skin.**—The body temperature can be approximately estimated by the hand placed first on the chest and then gently pushed into the axilla. Elevated body temperature is the principal symptom of fever. If the temperature is above 98.4° F. (37° C.), a differential diagnosis of the febrile diseases must be made. See Chapter II.

The most striking symptom, as a rule, to call attention to a febrile condition, is the flushed face. While the temperature is being taken, the pulse should be felt, and search for an eruption should be made.

If the patient is perspiring or the hand of the physician is cold, estimation of body temperature by the hand is unreliable.

11. Dryness of the skin and perspiration can either one be of diagnostic value. Great dryness is present in conditions in which there is a large amount of watery excretion, such as polyuria from any cause, diabetes mellitus, cholera, severe diarrhœa. Perspiration, also, may be of diagnostic importance. In some of the febrile diseases profuse sweating may indicate the approaching crisis. In chronic diseases it may be a sign of weakness, as in the night-sweats of phthisis. It often accompanies collapse and approaching death. Among drugs, the antipyretics of the coal-tar series, especially, occasionally produce profuse sweating. It may appear, however, now and then, in healthy persons, especially in the young, after a large ingestion of fluids.

12. The pulse. — By long-established precedent, the examination of the patient is begun by feeling of his pulse. By this procedure one can recognize: *a.* The presence or absence of fever. As a rule, the pulse-rate is quickened (over 90) in the febrile state; the tension is increased while the artery remains soft. *b.* The patient's general condition. A strong, healthy man has a full pulse of good tension; a person weakened by disease has a rapid, small pulse, with little tension. *c.* The presence of particular changes in the heart, or in certain other organs (Chap. VII.).

The pulse is felt by placing the finger (usually the index finger, *never* the thumb) upon the radial artery, a little above the wrist. The hand of the patient should not be raised from the bed. The beginner should accustom himself to count the pulse, watch in hand, for a quarter of a minute, and to determine the minute-rate from this calculation.

The feeling of the pulse is an art which can be learned only by constant and by long practice. Experienced physicians sometimes bring it to an extraordinary degree of perfection. Indeed, some of the most valuable diagnostic hints may be gained by this procedure. The older physicians, who were masters of observation, were accustomed to lay the greatest importance upon this practice.

13. Striking symptoms. — As a training in diagnostics, it is a good practice for the physician, after the completed general examination, to ask himself if he has omitted to take note of any striking symptom or symptoms. It is, of course, a matter of experience to observe and appreciate certain signs and symptoms which might easily be overlooked in the general examination, and would come to light only in the systematic examination of each organ.

Any single element in a patient's complex of symptoms may appear as a striking symptom, as, for instance, extreme pallor, dyspnœa and cyanosis or œdema. Certain diseases may produce striking symptoms, the observation of which may be the starting-point of the diagnosis; *e.g.*, *ascites* (fluid in the abdominal cavity), *meteorism* (distention of the abdomen through gas in the intestines), *enlarged glands*, *varicose veins*, *vomiting*, *peculiarities* in the *urine* or *sputum*, etc., etc.

Mention must be made of a few striking symptoms which are of more value as to the judgment of the present condition of a patient than for the establishment of a differential diagnosis.

1. A sudden change in the condition of a patient, characterized by a small and frequent pulse, great difficulty in respiration, extreme pallor, coldness of the extremities, nose, and ears, with a rapid decline of the body temperature, is known as *collapse*. It is caused either by an internal hæmorrhage or by sudden heart failure in the crisis, recurrence, or even convalescence of the febrile diseases. In the last stage of typhoid fever and after diphtheria, collapse must be looked for. It sometimes follows a rapid sitting up in bed, too early or too long absence from bed, or over-great exertion at stool; sometimes it appears without any appreciable cause.

2. An accumulation of fluids in the larger air-passages, which gives rise to the well-known *stertor* or *death-rattle*.

is a symptom of approaching death. It can be heard at inspiration and expiration, and at some distance.

3. What is commonly called the *death-struggle* or *agony* is the appearance of the progressive and complete paralysis of all muscular and nervous functions. It is characterized by unconsciousness, the facies Hippocratica, and the phenomenon of the death-rattle.

The *certain* signs of death are: cessation of respiration, of the sounds of the heart, of the pulse; absence of all reflexes, particularly that of the cornea.

It is certainly of rare occurrence that a physician is in doubt whether a seeming corpse is really dead. Should such a suspicion arise in the case of a sudden death, certain experiments will readily decide the matter; such as, the exposure and incision of an artery, electric stimulation of the muscles, or the insertion of a needle into the heart, application of down to the lips.

CHAPTER II

THE DIAGNOSIS OF THE ACUTE FEBRILE AND ACUTE INFECTIOUS DISEASES

In addition to the general considerations, the history, in a febrile case, must embrace: *previous infectious diseases* (typhoid fever, measles, and scarlatina attack an individual but once, as a rule, but pneumonia, erysipelas, and articular rheumatism may occur repeatedly); *direct causes of infection*, such as similar illness in the neighborhood, or exceptional opportunity for infection by food, water, etc.; *predisposing causes* (cold, emotion, trauma, errors of diet); *initial symptoms:* chill, headache, lassitude, sore throat, vomiting, pain in side or back, etc.

The febrile diseases in general can be recognized by the symptom complex of *fever*. This is characterized by a red, flushed face often covered with perspiration; quickened breathing; full, rapid, but soft pulse; pronounced thirst; diminished appetite; diminished quantity of urine of high specific gravity and dark-colored; *increased* body temperature.

The temperature should be taken at once with the *clinical thermometer*. The thermometer should remain in the axilla 10 minutes, in the rectum 5 minutes.

The temperature in the mouth is from 0.2° to 0.3° higher than in the axilla; the temperature taken in the rectum is about 1° F. (0.5°-0.8° C.) higher than that in the mouth. The temperature of a healthy person estimated in the mouth or rectum varies between 98.5° and 99.5° F. (36.9°-37.4° C.). It is lowest in the morning, and tends to rise in the evening from 0.5° to 1° F. or C. The Germans employ the scale of Celsius, the French that of Réamur, the Americans and English the Fahrenheit scale. The equivalent table is here given:—

$n°$ C. $= \frac{4}{5} n°$ R. $= \frac{9}{5} n° + 32°$ F.

C.		R.		F.
36°	=	28.5°	=	96.8°
37°	=	29.6°	=	98.6°
38°	=	30.4°	=	100.4°
39°	=	31.2°	=	102.2°
40°	=	32°	=	104°
41°	=	32.8°	=	105.8°

The so-called *self-registering thermometers* may be recommended. After the taking of the temperature, a little metal rod lying above the mercury shows the degree of fever until shaken down.

The *minute thermometers* are handy for general use. Because of their smallness and of a peculiar amalgam of mercury, they give the accurate axillary temperature in two minutes.

The temperature of healthy persons in the axilla is from 36.5° to 37.5° C. It is lowest in the morning, in the evening from 0.5° to 1.0° higher. Slight elevations of temperature appear *transitorily* after a heavy meal (digestive fever), great exertions, continued exposure to the sun (insolation), warm baths. Continued elevation of temperature is a sign of *fever*. Temperature below 36° C. is called *collapse* temperature; 36° to 37° C. normal; 37.5° to 38.0° C. *subfebrile* temperature; 38° to 38.5° C. *light* fever; 38.5° to 39.5° C. (evenings), *moderate* fever; 39.5° to 40.5° C., *high* fever; above 41.5° C., *hyperpyretic* temperatures.

When fever exists, the temperature varies during the day; in the morning there is a moderate decrease (*remission*), in the evening an increase (*exacerbation*). If the exacerbation comes in the morning, the remission in the evening, an *inverted type* of fever is present (most frequently in phthisis).

Chills: If the bodily temperature rises very suddenly while the radiation of heat is diminished by contraction of the cutaneous vessels, the patient has the sensation of intense cold which manifests itself in involuntary shivering, chattering of the teeth, shaking of the entire body. A chill appears: 1. in a single attack at the beginning of the acute infectious diseases (pneumonia, erysipelas, scarlatina, etc.); 2. in repeated attacks: (*a*) at regular intervals: in malaria (which may be suppressed by quinine), (*b*) at irregular intervals (uninfluenced by quinine), in the presence of deep abscesses and pyæmia, more rarely tuberculosis and endocarditis. — Chills in the course of a typhoid fever may indicate a relapse or one of many threatening complications, as intestinal hæmorrhage, perforation, venous thrombosis, emboli of the lung, etc., although *they may be without any significance*, probably through irritation of the intestinal ulcer.

For the especial diagnosis of the febrile diseases, it is necessary to recognize the *type and the course* of the fever; for this purpose, during the entire time of the presence of fever, the temperature is taken at regular intervals and arranged on a fever chart (see below); by this means the *fever curve* is obtained. Many of the acute febrile diseases have characteristic curves.

The *type* of the fever is ascertained by the differences between the morning and evening temperatures.[1] These may be: *continuous* fever in which the day's difference is not more than 1°; *remittent* fever, with a daily difference of more than 1°; *intermittent* fever, in which the fever lasts only a few hours, with an absence of fever during the rest of the day (attack of fever and interval of freedom from fever).

In the course of almost all the febrile diseases, three stages may be distinguished: the time during which the temperature continues to rise (*stadium incrementi*); the *fastigium*, or acme, the period of little change in the temperature, which is usually high; *defervescence* (*stadium decrementi*), the period of decreasing temperature. The decrease may come rapidly in a few hours: *crisis*. The crisis is often heralded by a diminution of the frequency of the pulse and an outbreak of perspiration[2]; or it may be preceded by a short, very high rise of temperature occasionally accompanied by delirium (*perturbatio critica*); not rarely the crisis is followed by (*epicritical*) delirium; sometimes collapse ensues. The slow decrease of fever extending over days is known as *lysis*.

[1] In practice, the difference between the highest and lowest temperatures taken on the same day; the temperature is usually taken, for one reason or another, more than twice daily.

[2] The sweat of the crisis is frequently of peculiar, not disagreeable, odor.

The course of the acute infectious diseases, especially those characterized by exanthemata, is divided as follows: (1) stage of *incubation:* from the time of contagion to the beginning of the phenomena of the disease; (2) *prodromal* stage: from the beginning of the fever to the breaking out of the eruption; (3) *eruptive* stage; (4) stage of *desquamation* or defervescence.

The character of the fever. In severe fever, febris *stupida* and febris *versatilis* are distinguished. The former is marked by apathy, a dull, listless countenance, absolute quiet; the latter by an anxious expression, jactation, a mild delirium, picking at the bed-clothes. The transition from the former to the latter state is of evil significance.

Pathognomonic symptoms. After the reading of the temperature, or after the recognition of the fever curve, other quickly appreciable signs which aid in establishing a diagnosis must be looked for. It is well to accustom oneself to a certain routine. (It is best first to inspect the face and skin, then the other organs from above downward.)

1. *Exanthemata.* Characteristic eruptions appear in measles, scarlatina, typhoid fever, typhus fever, variola, varicella, erysipelas. The eruption decides the diagnosis. The exanthemata do not always appear at the beginning of the fever and often disappear before the fever, so that their aid must sometimes be foregone.

2. *Involvement of the sensorium.* Deep apathy is characteristic of typhoid, meningitis, miliary tuberculosis, severe forms of sepsis, and ulcerative endocarditis. Delirium is of no value for the purposes of differential diagnosis.

3. *Herpes labialis and nasalis* (small blebs with serous contents, at angle of the mouth and on the nose, which soon dry up and leave a brownish crust). Herpes is very common in epidemic meningitis and pneumonia, and speaks against tuberculosis, typhoid, and pleurisy.

4. *Frequency of the pulse* under certain circumstances may be of great diagnostic importance. Very slow at the

beginning of meningitis. In scarlatina unusually rapid. In typhoid, aids in the diagnosis of the stage of the disease; in the first stage, comparatively slow, usually not over 110 in uncomplicated typhoid; in the third stage usually from 100 to 120.

5. *Involvement of other organs.* *Lips:* fuliginous (sooty color) in typhoid. *Tongue:* raspberry-like in scarlatina, coated in typhoid. *Neck:* characteristic affections in angina and diphtheria. *Stiffness of the neck* in meningitis. *Rusty sputum* in pneumonia. *Distended abdomen* painless to the touch, in typhoid; retracted abdomen in meningitis. *Enlargement of the spleen* particularly important in typhoid and malaria (Chap. IV.). *Diarrhœa* of characteristic type in typhoid, dysentery, cholera. Redness and swelling of several *joints* in acute articular rheumatism. Condition of the *urine:* diazoreaction of the *urine* in typhoid, etc. Condition of the *blood:* in the majority of the acute infectious diseases the number of the white blood cells is increased (infectious *hyperleucocytosis*); this is absent in typhoid, malaria, glanders, and in many cases of septicæmia.

In many cases it will be possible to make an early diagnosis of the infectious disease present from the recognition of the type of the fever and the consideration of the general and special symptoms.

Yet it must not be forgotten that several days' observation is required for an insight into the course of the fever and that many characteristic symptoms do not develop until the disease is well under way (*e.g.*, exanthemata, enlargement of the spleen, the diazoreaction, diarrhœa, etc.). It is often necessary, therefore, judging from the temperature and the habitus of the patient, to content oneself with the temporary diagnosis "acute infectious disease" and to institute the necessary general therapeutic measures (rest in bed, comfortable position, light bed-clothing, cool fluid diet, ice-bag to the head, diluted acids, trained nursing). These therapeutic directions are independent of any special diagnosis for a few days, when it is usually possible to make a differential diagnosis from the developed symptoms.

Symptoms of the Acute Infectious Diseases

I. Acute Infectious Diseases with Regular Course of Fever

Measles (*morbilli*) (Fig. 1). — Incubation 10 days, with coryza, cough, gastric disturbances. Prodromata, 2 to 3 days, beginning with chills and high fever. On the 2d or 3d day, diminished fever; on the 3d or 4th day, appearance of the measles eruption, with high temperature. From the 4th to 7th days, continuous fever; on the 7th day, crisis, sometimes lysis. Desquamation for about 14 days, in small scales (furfuraceous desquamation).

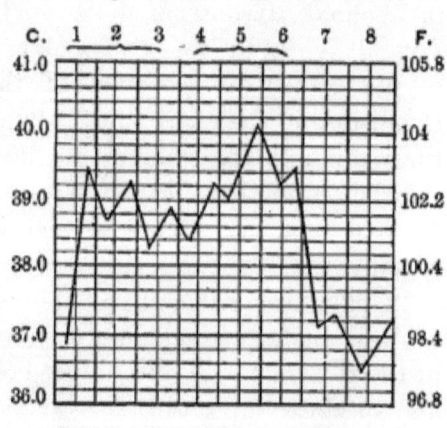

Fig. 1. — Fever Curve in Measles.

Further principal symptoms: coryza, cough, conjunctivitis, and photophobia. Pulse somewhat accelerated (in children from 140 to 160). Rare but dangerous complication: broncho-pneumonia. The rare complications with diphtheria and croup suggest a grave prognosis.

Scarlet fever (scarlatina) (Fig. 2). — Incubation, 2 to 24 days, usually without symptoms; prodromata, 1 to 2 days, beginning with chill and high fever. On the 2d day, the eruption of scarlet-red color, with rising temperature. From the 4th day, lysis. Desquamation, 4 to 14 days; large scales sometimes fall off (*desquamatio membranacea*). Sometimes the fever of scarlatina is in no way typical.

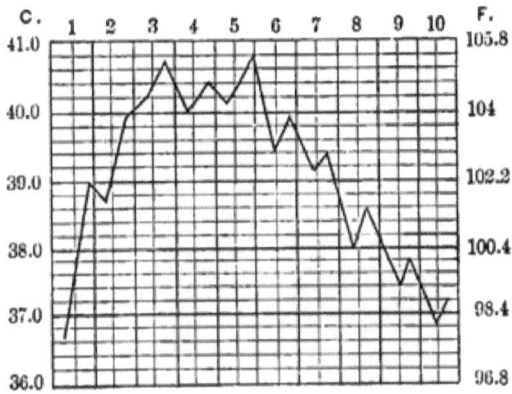

Fig. 2.— Fever Curve in Scarlatina.

Further important symptoms: *angina*, raspberry tongue, vomiting during the prodromal stage. *Complications and sequelæ:* acute nephritis, otitis, more rarely glandular suppuration, joint affections (usually benign), endocarditis.

Erysipelas (Fig. 3). — Incubation, 1 to 8 days. Begins with chill and high fever. On the 1st and 2d days, redness and swelling of the skin. Continuous fever during the extension of the erysipelas. The redness and swelling frequently appear in batches, calling forth irregular remittent or intermittent fever.

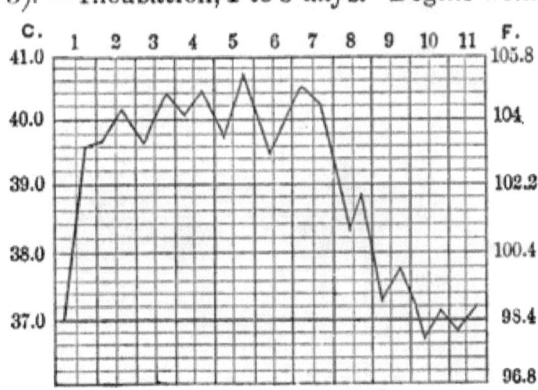

Fig. 3.— Fever Curve in Erysipelas.

The redness and swelling are often confined to the face or to the hairy part of the head (*erysipelas capitis et faciei*), but may extend to the neck and trunk. Erysipelas may follow an injury in any part of the body.

Croupous pneumonia (Figs. 4 and 5). — Begins with chill, high fever, and pain in one side of the chest. Continuous fever. Crisis between the 3d and the 11th days, often on an odd day of the disease, most frequently on the 5th and 7th. Crisis on the 3d day deceptive (Fig. 5); usually followed by renewed continuous fever. Sometimes the crisis is extended over several days (protracted crisis).

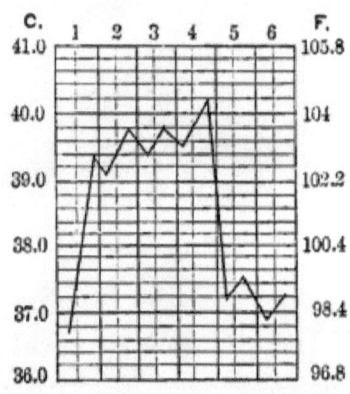

FIG. 4. — FEVER CURVE IN PNEUMONIA. CRISIS 4TH TO 5TH DAY.

Pathognomonic sign: rusty sputum, although some patients with pneumonia have no expectoration, or a whitish-yellow sputum which is not characteristic.

Enlargement of the spleen, which does not disappear until complete resolution has taken place.

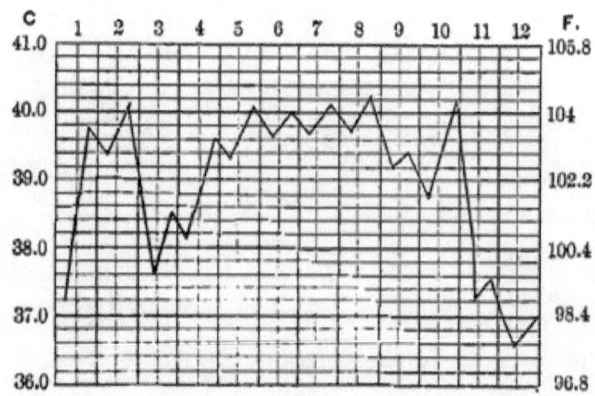

FIG. 5. — FEVER CURVE IN PNEUMONIA. PSEUDOCRISIS ON THE 3D, CRISIS ON THE 11TH DAY.

A new rise of temperature, indicating an advancing infiltration, may follow the crisis. This is called *pseudocrisis*. This is to be diagnosticated, when in spite of the fall of temperature, the pulse and rapidity of the respiration remain abnormally high; or when, despite the normal temperature, the hyperleucocytosis persists.

Physical signs of the infiltrated lung at the height of the disease: dulness (with a tympanitic note) and bronchial breathing (cf. Chap. VI.).

If a regular crisis does not supervene, or if the fever rises irregularly after the crisis, it is to be referred to a pleurisy (serous or purulent), more rarely gangrene, tuberculosis, or abscess of the lung.

Typhoid fever (Fig. 6). — Incubation 7 to 21 days. Prodromata for about a week, with indefinite symptoms of

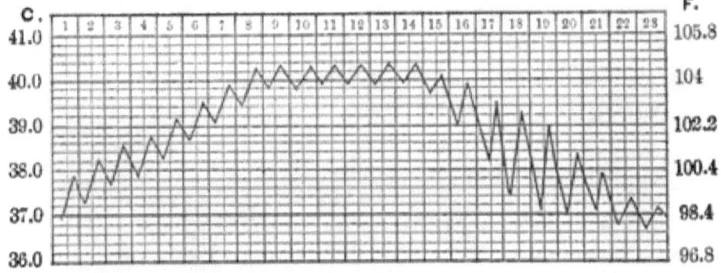

Fig. 6.—Schematic Fever Curve in Typhoid Fever.

general lassitude. *Stadium incrementi*, terrace-like rise of temperature of remittent character as it ascends. Acme reached on 4th to 7th day. *Fastigium*, continuous fever. *Defervescence*, fever remitting as it descends. Morning temperatures sink daily; during the first days the evening temperatures reach a high point (uncertain stage, steep fever curve).—The duration of the 2d stage varies with the severity of the disease. In light cases the descending remission of the fever may begin on the 16th or 14th days or even earlier; in very severe cases the continuous fever may last until the 5th week.

Further principal symptoms: apathy, stupor, fuliginous, furred tongue. Roseola from the end of the 1st to the middle of the 2d week. Enlargement of the spleen during the acme. Meteorism. Stools, diarrhœal of pea-soup consistency. Diazoreaction in the urine, no hyperleucocytosis. Frequently, bronchitis.

The diagnosis is made by the presence of several of these signs at the same time; it is scarcely possible to do so from any single one. Many symptoms may be absent; for instance, decided meteorism, the diazoreaction, the diarrhœa.

The diagnosis of typhoid must embrace the week and stage of the disease, if possible. This is sometimes facilitated by complications which mark certain periods; thus, *intestinal hæmorrhages* take place most frequently during the time of defervescence; and during this period, too, fatal perforation of the intestine sometimes occurs (vid. p. 101).

The prognosis depends, among other things, upon the frequency and tension of the pulse and the degree of stupor.

Typhus fever (*typhus exanthematicus*, spotted fever) (Fig. 7). — Incubation 3 to 21 days. No prodromata. Begins

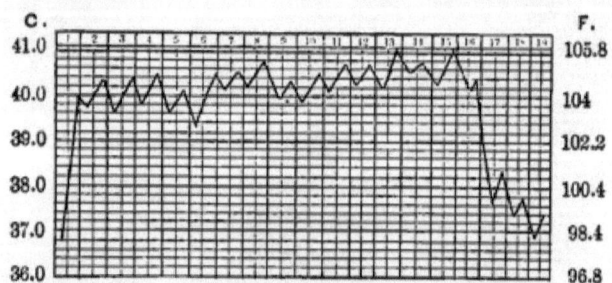

FIG. 7. — FEVER CURVE IN TYPHUS FEVER.

with chill and high temperature. On the 3d day, a profuse roseola. Continuous fever from 13 to 17 days with slight remissions on the 6th to the 8th day. Temperature falls by crisis, with delirium.

The roseola soon becomes petechial. Bronchitis. The mental symptoms especially severe.

Recurrent fever (relapsing fever) (Fig. 8). — Incubation 5 to 7 days. No distinct prodromal stage. Begins with chill and rapid rise of temperature. Continuous fever 5 to 7 days. Temperature falls by crisis. Then 5 to 8 days' absence

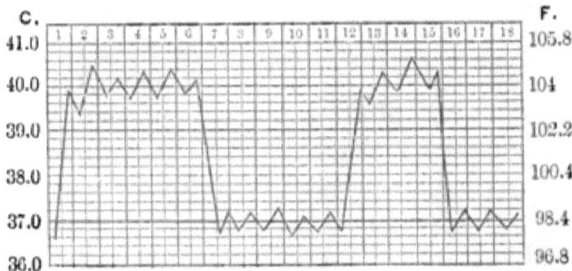

Fig. 8.—Fever Curve in Relapsing Fever.

of fever; thereupon renewed continuous fever, usually of short duration. Frequently after absence of fever for 7 days, continuous fever for from 2 to 3 days. — Great rapidity of pulse. Enlargement of the spleen. Roseola. Herpes. *Spirochetæ Obermeieri* in the blood during the fever (Chap. XII.).

Variola (small-pox) (Fig. 9). — Incubation 10 to 13 days. The actual course of the disease may be divided into 4

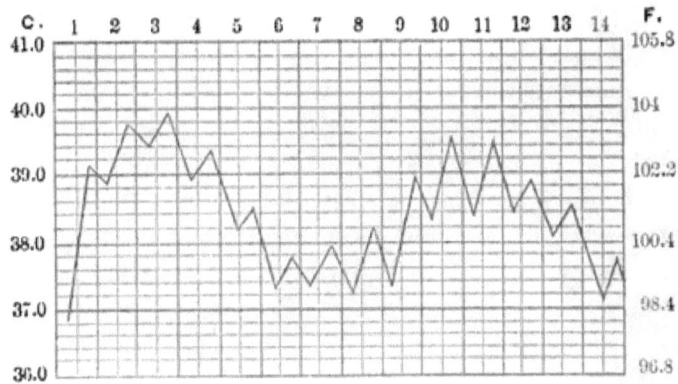

Fig. 9.—Fever Curve in Variola.

stages *stage of invasion* (prodromata), beginning with chill and high fever, 3 to 4 days. *Stage of eruption*, diminished fever to the 9th day. *Stage of suppuration* (fever of pus), severe remittent fever, 9 to 11 days. *Stage of exsiccation* (drying up of the papules), fall of the temperature by lysis.

The eruption forms at first red spots which gradually change into larger papules. On the 6th day they become filled with a cloudy fluid, on the 8th day they are blebs filled with pus; from the 9th day the contents are poured out, and the drying up begins on the 11th day. The eruption attacks the mucous membrane of the mouth and throat. — Severe drawing pains in the knees and back.

Varioloid (Fig. 10) is the mildest form of small-pox as it is seen, after contagion, in persons who were unsuccessfully vaccinated or vaccinated more than ten years previously. In varioloid, the period of invasion is followed immediately by the stage of drying up, without the fever of suppuration.

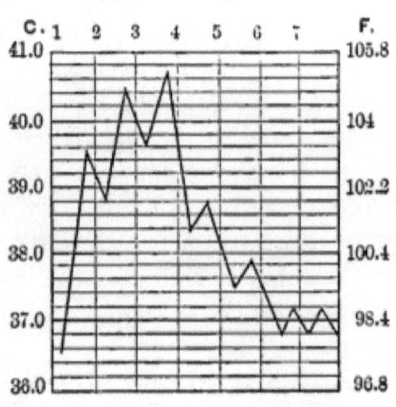

FIG. 10.—FEVER CURVE IN VARIOLOID.

The eruption may be merely indicated or it may appear in some irregular form (erythema).

Varicella (chicken-pox). — Fever begins with a chill and is continuous until the drying up of the eruption, 2 to 4 days.

Eruption characteristic: rose-colored, slightly-elevated spots, which soon become vesicular. Found on the gums and throat also. Seldom like variola, but then to be distinguished by the fact that in varicella all the stages of the eruption appear simultaneously. Prognosis absolutely good. In very rare cases sequelæ are seen; for instance, acute nephritis.

Malaria (intermittent fever) (Figs. 11, 12, 13). — Incubation, 7 to 21 days. No distinct prodromata. Chill with high

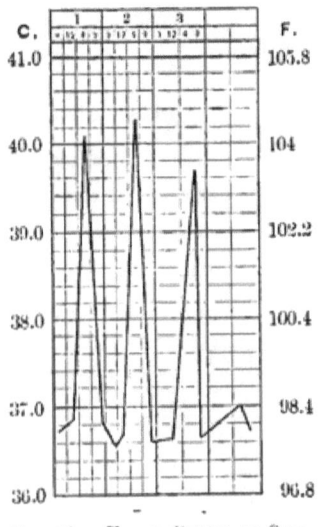

Fig. 11.—Fever Curve in Quotidian Intermittent Fever.

elevation of temperature; in a few hours, critical descent of fever, with perspiration, then apyrexia. Shortly before and during the fever, the cause of the malaria, the *plasmodium of Laveran*, in the blood. The attack is repeated at the same time of day the following day (quotidian type), Fig. 11; or on alternate days (tertian type), Fig. 12; or every fourth day (quartan type), Fig. 13. If the introductory chill appears before or after the regular time of day, it is spoken of as anticipated or postponed intermittent fever. Two attacks in one day characterize double intermittent fever.

The diagnosis of malaria is rendered absolutely certain in doubtful cases by the microscopic demonstration of the plasmodium in the blood; also by the establishment of the type of the fever, the enlargement of the spleen, and the specific controlling action of from 1 to 2 grammes of quinine when this dose is administered 6 hours before the expected attack of fever.

In the tropics there are forms of malaria in which the fever is entirely irregular; at the same time organic affections appear,

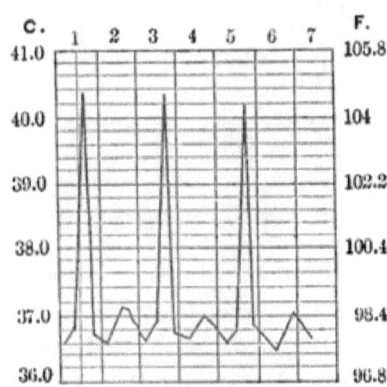

Fig. 12.—Fever Curve in Tertian Intermittent Fever.

which totally obscure the picture of the disease (masked intermittent fever). The diagnosis is made certain only by the curative action of quinine.

In the temperate climates, *malarial neuralgia* is seen, attacks of neural pain which return at certain times of the day like the paroxysms of fever. They are controlled by quinine (*e.g.*, supraorbital neuralgia).

Irregular fever, intermittent in type, which does not yield to quinine, and in which the malarial plasmodia can not be found in the blood, is to be referred to *deeply lying abscesses* or *endocarditis* or *latent tuberculosis*.

Influenza: after a short prodromal stage, sudden, usually high fever, which lasts for several days in continuous and remittent form, accompanied by intense prostration and great pain in the extremities. Frequently no localization, frequently catarrh of the bronchi. The unusually large number of complications or sequelæ is characteristic: respiratory tract (catarrhal and croupous pneumonia); circulatory apparatus (endocarditis, thrombosis); nervous system (neuralgias and psychoses), etc. The prognosis is good except when complications or sequelæ set in.

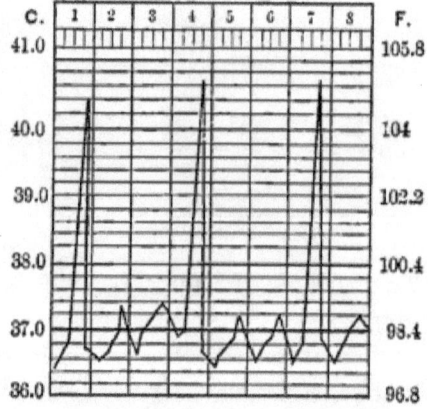

Fig. 13.—Fever Curve in Quartan Intermittent Fever.

II. Acute Infectious Diseases without Regular Course

A number of acute infectious diseases run their courses with irregular fever which does not conform to any type. The diagnosis of these diseases rests upon the local lesions.

Follicular amygdallitis (*angina follicularis*).—Redness and swelling of the soft palate, the tonsils, and the pharynx, often accompanied by a white or grayish exudate, which can usually be removed without causing hæmorrhage. Streptococci are most often found in the membrane. Submaxillary glands frequently involved. Fever, beginning with a chill, continuous for several days or lightly remittent. General disturbances not very severe despite the high temperature; if severe, diminished in a few days. An acute nephritis may appear after a light attack of angina.

Abscess of the tonsil sometimes supervenes (*angina apostematosa*).

Diphtheria. — Tonsils and palate covered with a grayish-white membrane, the removal of which produces bleeding from the mucous membrane. Diphtheria bacilli may be obtained from this membrane by culture. In severe cases, the membrane involves the nose, the larynx, the bronchi. Enlargement of the submaxillary glands. The profound general disturbance is characteristic (small, rapid pulse, stupor). Frequently, albuminuria. Fever throughout atypical, and gives no data for the prognosis, which depends in part upon the severity of the local affection, in part upon the profundity of the general infection. Frequent characteristic sequelæ: paralysis of accommodation, of the palate, of the extremities. There are cases of pharyngeal inflammation, in which the diagnosis between diphtheria and angina can not be made for some time; the decision must be reached by a bacteriological examination (cf. Chap. XII.).

Acute miliary tuberculosis. — Atypical fever. Marked cyanosis and dyspnœa. Over extensive areas of the lungs crepitant râles without dulness. In the urine, diazoreaction. In some cases tubercles in the choroid may be found ophthalmoscopically. Prognosis very bad; death in from 8 to 14 days.

Cerebro-spinal meningitis. — Irregular, partly remittent, partly continuous fever, with a long course and many remissions. Deep stupor. Pathognomonic symptom: stiffness of the neck. In the early stages, hyperæsthesia of the extremities, slow pulse.

The diagnosis must include the etiology. 1. *Epidemic* (sporadic) meningitis is diagnosticated by positive exclusion of other causes; during an epidemic the diagnosis is easier. 2. *Tubercular* meningitis may be diagnosticated in the presence of tuberculosis of the *lungs*, usually somewhat advanced. Herpes labialis never present. Recently this diagnosis has been repeatedly made by finding tubercle bacilli in the cerebro-spinal fluid obtained by puncture of the cerebro-spinal canal (lumbar puncture of *Quincke*). 3. Meningitis due to an extension of an *otitis media*.

The *stiffness of the neck* is usually not very distinct during the first few days, but is well elicited by the end of the first week. If the meningitis advances, the stiffness disappears, but the symptoms of paralysis supervene. A mild stiffness of the neck may be present at the beginning of severe cases of pneumonia and typhoid; meningitis may accompany these diseases.

Acute articular rheumatism.—Irregular remittent fever. Redness, swelling, and pain in several joints, usually affecting both sides of the body symmetrically. Fever and local affection yield in two-thirds of all cases to salicylic acid or antipyrin. Frequent complications: endocarditis, with permanent valvular lesions; more rarely, pleurisy, pericarditis. Atypical forms of articular rheumatism, especially in gonorrhœa.

Parotitis epidemica (mumps).—Irregular, moderately high fever, lasting from 2 to 8 days. Pain, enlargement of one or both parotids; resolution usual, sometimes suppuration. In a few cases, accompanied or followed by a febrile orchitis, more rarely epidydimitis.

Sepsis (pyæmia, septicæmia, blood poisoning).—Atypical, usually remittent fever, commonly with irregular chills. Stupor, great prostration. Usually no hyperleucocytosis. Primary suppuration often present, as, infected wounds, in the uterus (puerperal sepsis), in paronychia and phlegmonous processes, in cavities of the head (empyema of the antrum of Highmore, otitis media), in the prostate. The entrance of the producers of the process may remain hidden (cryptogenetic septicæmia). Shortly before death the responsible bacteria may be found in the blood. *Pyæmia* is characterized by localized abscesses in various organs of the body, while *septicæmia* denotes the general toxæmia without localization. These two forms of disease frequently pass over into each other. In the pure pyæmic type, hyperleucocytosis usually exists. Differentiation between the various kinds of sepsis according to the etiology (streptococci, staphylococci, diplococci, etc.) has no clinical value.

Acute endocarditis. — Irregular, usually remittent fever, with the physical signs of a lesion of a cardiac valve (intracardiac murmur; the systolic murmur alone does not denote an endocarditis). The diagnosis must differentiate between endocarditis of benign (verrucous or warty) and of malignant (ulcerative) character. The latter diagnosis is certain when repeated chills and many cutaneous emboli manifest themselves. A benign endocarditis frequently becomes malignant. An acute endocarditis develops during or after other infectious diseases. Endocarditis following articular rheumatism is usually warty, and disappears, leaving valvular lesions. Endocarditis accompanying sepsis, pneumonia, gonorrhœa, usually ulcerative.

CHAPTER III

DIAGNOSIS OF THE DISEASES OF THE NERVOUS SYSTEM

The following data are especially important in the anamnesis and history of nervous diseases: (1) *Heredity* in psychical diseases, neurasthenia, epilepsy, hysteria, syphilis in the parents. (2) Previous disease, especially syphilis, acute infectious disease. (3) Causes and predisposing factors: trauma, exposure, fright, intoxications (lead, mercury, alcohol, excessive use of tobacco).

The serial data for registering the *status præsens* are to be found on page 2, under B. I. The disturbances of sensation and motility require special consideration (B. I. 6), for which the following guide or scheme is in use at the 1st Medical Clinic of Berlin:—

STATUS OF NERVOUS SYSTEM

A. Motion

I. *Position of the limbs at rest.*
   a. Abnormalities in position.
   b. Atrophy?
   c. Abnormal involuntary movements (tremor, convulsive movements)?

II. *Movements.*
   a. Free, active.
   b. With resistance, active.
   c. Passive.

1. Face

I. Are both halves symmetrical?
   Equal palpebral fissures?
   Equal nostrils?
   Mouth, straight or crooked?
   Are the eyes normal or quiet?
   Pupils equal?

II. Request the patient to wrinkle the forehead, close the eyes, pucker the mouth, laugh, blow out the cheeks, extrude the tongue, move it to right and left, move the eyes (to the right, left, upwards, downwards, and to converge).

2. Cavity of the Mouth; the Larynx

I. Condition of the soft palate and uvula.
II. Elevation of the same in phonation.

3. Throat and Neck

I. Position of the head.
II. Turn head to right and left, bend it forwards and backwards (oppose resistance lastly).

4. Shoulders and Arms

I. Appearance of the scapula, position of the arms, of the fingers, size of the thenar and hypothenar eminences, condition of interosseous spaces.
II. Elevate the shoulders, raise the arms (to the vertical position), abduct them, hold them horizontally forwards.
Flex and extend the forearm.
Pronate and supinate.
Flex and extend the fingers, spread them apart, close the thumbs upon the palms.
Force of grasp.

5. Legs

I. Condition of the trochanters.
II. Elevation, abduction and adduction.
Flex and extend the legs.
Flex and extend the feet.

6. Trunk

I. Breathing (rhythm)?
Abdomen (retracted)?
Spinal column (shape)?
Buttocks (hypertrophied or atrophied)?
II. (Only necessary if a disturbance is perceived during inspection).
Bend trunk forwards, backwards, to the sides, take deep inspiration, cough.

7. **Bladder and Rectum** (inquiry)
8. **Equilibrium with closed eyes**
9. **Gait**
10. **Ability to grasp objects**
11. **Speech**
12. **Chirography**

B. SENSATION

I. *Subjective signs:* deafness, formication, pain, etc.
II. *Objective examination.*
   a. Sensibility of the skin.
      1. Most gentle contact.
      2. Painful pin pricks.
      3. Cold.
      4. Heat.
   b. Muscular sense.
      1. Sensation of movement.
   c. Visual field.
   d. Hearing.
   e. Smell.
   f. Taste.

C. REFLEXES

a. Skin reflex.
Reflex of sole and palm, cremaster, abdominal, lids, conjunctiva, palate.
b. Tendon and periosteal reflex.
Patella, tendo Achillis, foot clonus, wrist, biceps, triceps.
c. Pupillary reflex.

Condition of the sensorium: dulness, stupor, apathy in *febrile* diseases is an indication of an intense infection, and suggests a grave prognosis. These mental states are of diagnostic value in typhoid, meningitis, miliary tuberculosis, sepsis.

In *non-febrile* diseases apathy and stupor are, as a rule, the precursors of complete unconsciousness (*coma, lethargy*).

Coma is present in cases of: 1. Poisoning; in these cases the history will determine the condition, or, better, the examination for poison of the contents of the stomach obtained by lavage. 2. Apoplexy; often recognized by the coincident facial palsy, or by a hemiplegia; the deeper and the longer in duration the coma, the worse the prognosis. 3. In increased cerebral pressure, for example, induced by tumors, associated with choked disc. 4. In the course of the diseases due to perverted or defective metabolism, especially diabetes; recognized by the deep and laborious respiration (intense breathing of *Kussmaul*). Usually lethal in termination. It is seldom present in carcinoma and anæmia. 5. During the course of nephritis (*uræmic coma*), commonly preceded by repeated attacks of convulsions and high arterial tension. An important diagnostic rule may be thus derived: in every case of coma arising from an unknown cause, the urine should be carefully drawn by catheter and examined for both *albumin* and *sugar*. Markedly rigid radial arteries would point to cerebral apoplexy. In many cases the diagnosis will be determined by the history obtained from the relatives and friends.

Dulness of the sensorium, ranging from slight apathy to stupor, is often a sign of an existing *psychosis*, which could be diagnosed by the presence of hallucinations and delusions, but better by the exclusion of all the above-mentioned visceral causes of coma.

It is especially important for the diagnostician to recognize the *psychoses* due to certain drugs (salicylic acid, bromides, etc.), to chorea, to certain of the last stages of heart affections, to all conditions of inanition; in all these there are illusions and hallucinations of hearing and sight, as well as delusions of persecution and those of a religious character.

Essential mental disturbances (mania, melancholia, paranoia, paretic dementia) in well-developed cases are readily recognized, and would most likely be turned over to asylums (strict legal forms being observed). Most frequently the first indistinct beginning of dulness of judgment or of perception succumbs to the acumen of the physician.

To determine such indefinite psychical anomalies (motiveless pleasurable and painful feelings, sudden change of mood, diminution of intelligence and of memory, moral decrepitude) the physician must carefully investigate whether the case is one of psychic reaction to a somatic functional disturbance or of a beginning insanity. In the latter case the signs of an impending paralytic dementia ought to be looked for: iridoplegia, inequality of the pupils, absence of or increase of the knee jerk, disturbances of speech, etc. The further elucidation of this most highly important question would be transferred to the psychiatrist.

Delirium in febrile diseases is conditional to the height of the fever and of the intoxication; it is without essential prognostic significance. It occurs frequently before the crisis of the disease (*perturbatio critica*), after its crisis (delirium of defervescence), and it is often the sign of collapse. When accompanying non-febrile affections (irrespective of the psychoses and intoxications, especially alcoholism), it is always an indication of intense exhaustion (inanition, delirium).

Diffuse headache when *transitory* is usually of small diagnostic importance (as in fever, over-exertion, excesses, dyspepsia, constipation, female genital affections, etc.). *Paroxysmal* headache is often confined to *one side* of the head. *Migraine:* the latter is chiefly associated with dyspeptic signs, and often terminates in an attack of vomiting; not infrequently is it limited to the course of distribution of certain nerve trunks; when occurring thus, painful points on pressure may be elicited. *Neuralgia:* continuous headaches occur in neurasthenia as well as in chlorosis, heart diseases, etc.; when occurring with diabetes it denotes a grave form of the disease. *It is often the first sign of a developing uræmia.* It also occurs in tertiary syphilis, and is then especially nocturnal (*dolores osteocopi*).

Vertigo. Occurs frequently in gastric and intestinal diseases. When present in a heart affection it points chiefly to a stenotic lesion. It accompanies anæmia. When concomitant with shrill tinnitus aurium, it indicates an aural disease (labyrinth), and is known as the *Ménière* symptom. It is also present in cases of cerebral tumors and special affections of the cerebellum. It is not infrequently associated with imperative movements.

Anatomical Introduction

An intimate knowledge of the anatomy of the central nervous system is essential in order to form a correct diagnosis of the various forms of disease of the nervous system.

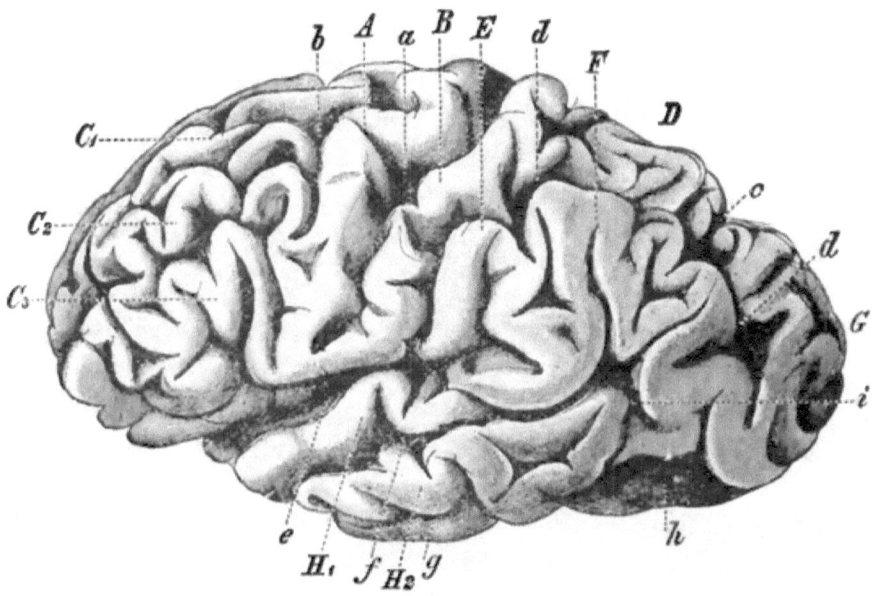

Fig. 14.—External Surface of the Left Cerebral Hemisphere.
(*Drawn from a brain treated with nitric acid and dried*)

a	central fissure	B	post-central gyrus
b	pre-central fissure	C_1	superior frontal gyrus
c	parieto-occipital fissure	C_2	middle frontal gyrus
d	intraparietal fissure	C_3	inferior frontal gyrus
e	Sylvian fissure	D	parietal lobe
f	superficial temporal fissure	E	marginal gyrus
g	second temporal fissure	F	angular gyrus
h	inferior occipital fissure	G	occipital lobe
i	anterior occipital fissure	H_1	first temporal gyrus
A	pre-central gyrus	H_2	second temporal gyrus

The most important anatomical features which are necessary for clinical work are here briefly presented.

The **motor** tracts proceed from the psychomotor centres of the cerebral cortex (Fig. 14).

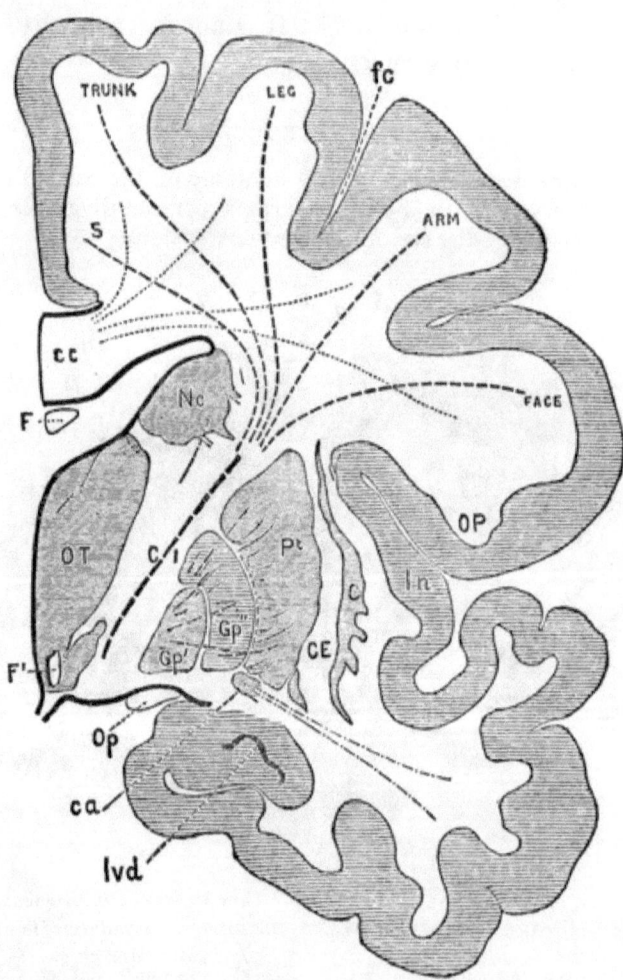

Fig. 15.—Outline of a Transverse Dorso-ventral Section of the Right Half of the Brain. (Natural size.) (Sherrington.)

OT, optic thalamus; *Nc*, nucleus caudatus; the head only appears in this section; *Pt*, putamen; *Gp″*, *Gp′*, the two parts of the globus pallidus of the nucleus lenticularis; *C*, the claustrum; *CE*, the external capsule; *In*, the island of Reil; *ca*, the anterior commissure shaded to render it distinct, and the fibres from the temporo-sphenoidal lobe which pass into it being indicated by broken lines; *Op*, the optic tract; *lvd*, the end of the descending horn of the lateral ventricle; *F*, the fornix; *F′*, the end of the anterior pillar of the fornix in the base of the thalamus; *cc*, corpus callosum; *OP*, anterior part of the occipital lobe.

fc is the central fissure or fissure of Rolando. The course of the fibres of the pyramidal tract connected respectively with the trunk, leg, and arm, and hence with spinal nerves, and of those connected with the face and hence with cranial nerves, is shown by broken lines. These are all seen converging into the internal capsule, *CI*.

The centre for the voluntary movements of the arm lies in the middle third of the pre-central convolution; the centre for the face and tongue in the lower third of the same convolution; that for the motion of the legs, in the upper third of both the pre-central and post-central gyri and in the para-central lobule, which unites these two convolutions on the mesal face of the hemisphere. The motor centres of speech lie in the posterior part of the left inferior (third) frontal gyrus and in the insula (island of Reil) (between the inferior frontal and the superior temporal gyri). The sensory speech centre lies in the temporal lobe (first left temporal gyrus). The cortical visual centre is situated in the occipital lobe (for hemianopsia see p. 38, cerebral nerves), the cortical auditory centre in the temporal lobe.

The motor fibres proceed from the cortical centres through the *corona radiata* to the *internal capsule* (Fig. 15).

Here will be found the pyramidal (motor) tract in the middle third of the posterior half of the internal capsule, and between the thalamus and lenticular nucleus, closely bordering on the fibres which control voluntary facial movements. This location is the seat of predilection for cerebral hæmorrhages.

Proceeding through the internal capsule, the motor fibres continue to the pes (ventral portion of crus cerebri); the sensory fibres proceed through the tegmentum (dorsal portion of crus cerebri) into the pons. From the *pons* they go into the *medulla (oblongata)*, where they constitute the *pyramids*, and there decussate for the most part. In the *spinal cord* (Fig. 16) the decussated motor fibres proceed downwards in the *crossed pyramidal tract*, the very few fibres which did not decussate proceeding downwards as the *uncrossed* (anterior) *pyramidal tract*.

The motor fibres proceed from the pyramidal tracts to the *anterior cornua* of the central gray matter of the spinal cord, and from there leave the cord by way of the *anterior spinal roots* and the peripheral nerves to reach the muscles.

The trophic centres for the pyramidal tract lie in the cerebrum, and when any part of the motor tract is injured there ensues, in addition to the consecutive paralysis, a descending degeneration of the pyramidal tract; the trophic centres for the peripheral motor nerves lie in the ganglion cells of the anterior cornua. A lesion of this organ or one peripheral to it induces degeneration of the nerves and paralysis and atrophy of the muscles supplied by the structures concerned.

The sensory tracts run upwards in the *posterior columns* and posterior cornua of the cord; they decussate immediately after their entry and also in the lemniscus.

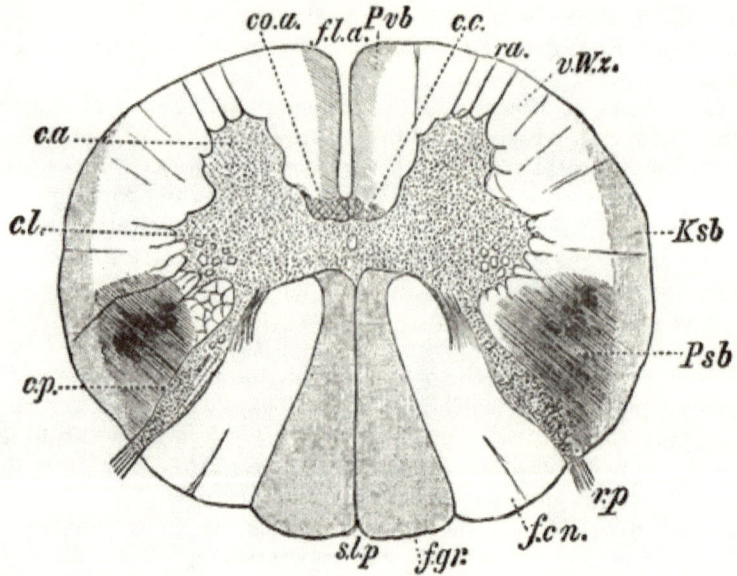

Fig. 16. — Diagrammatic Cross-section of the Spinal Cord showing the Tracts of the White Matter.

c.a. (*cornu anterius*) anterior horn
c.l. (*cornu laterale*) lateral horn
c.p. (*cornu posterius*) posterior horn
c.c. central canal in the gray commissure
co.a. anterior white commissure
r.a. (*radix anterior*) anterior root
r.p. (*radix posterior*) posterior root
f.l.a. (*fissura longitudinalis anterior*) anterior fissure
s.l.p. (*septum longitudinale posterius*) posterior septum
f.gr. (*funiculus gracilis*) Goll's column
f.cn. (*funiculus cuneatus*) Burdach's column
Ksb (*Kleinhirnseitenbahn*) direct cerebellar tract
Psb (*Pyramidalseitenbahn*) crossed pyramidal tract
Pvb (*Pyramidalvorderbahn*) direct pyramidal tract
v.Wz. (*Vordere Wurzelzone*) anterior radicular zone

PARALYSIS

(The absolute inability to move a limb is called *paralysis*; to motor weakness of a limb is given the name *paresis*.)

When the presence of a paralysis or paresis is determined, the following features should be elicited: —

1. Is the paralysis limited to one side of the body (*hemiplegia*, for example, paralysis of the right arm and

right leg); does it involve both sides (*paraplegia, e.g.*, both arms or both legs); or is it limited to one extremity or to a group of muscles (*monoplegia, e.g.*, one arm, or the servati group)? All hemiplegias are produced by lesions of the cerebrum, paraplegias by lesions of the spinal cord, although multiple neuritis occasionally is the cause of the latter. Monoplegias depend upon lesions either of the cerebrum or of the peripheral nerves.

2. Is the affected member *flaccid* or *spastic*?

Flaccid paralyzed limbs are easy to move by passive motion, but a spastic condition of the limb presupposes an extensive resistance to any attempts of motion, or a convulsive contraction ensues on such attempt. The reflexes in the latter condition of spastic paralysis are increased in the affected limb.

Spastic conditions ensue when contracture develops in the paralyzed muscles or in their antagonists, and further in those conditions of increased reflexes like those accompanying degeneration of the pyramidal tract.

3. Does *atrophy* exist? Atrophic palsies depend upon lesions of the peripheral nerves or of the anterior cornua of the spinal cord, or of the cranial nerve nuclei or of the pons.

The organic paralyses are in direct contrast to the *functional*, which do not depend upon anatomical causes but upon an affection of volition (hysterical paralysis, paralysis of fright). The functional palsies are recognized by the following criteria: there is an absence of trophic and electrical disturbances; there are present and coincident with the palsy certain definite signs of hysteria, such as hemianæsthesia, convulsions, contractures, etc.; above all they may be recognized by the general psychical character of the patient.

Hemiplegia

It is important in this condition to determine the etiology and to locate the part of the cerebrum which is involved in the process producing this symptom (localization).

Etiology

1. *Embolism.* The hemiplegia is suddenly produced, involves chiefly the Sylvian artery, and is thereby associated with motor aphasia. It is necessary to determine the origin of the source of the embolism, *e.g.*, affections of the left side of the heart.

2. *Apoplexy.* It appears suddenly, and is proven by an existing arterio-sclerosis or of contracted kidney. It involves chiefly the large central ganglia (corpus striatum, internal capsule) (Fig. 15), and occurs frequently without aphasia. It is usually accompanied by a long-continued comatose condition.

3. *Syphilis.* It may be the result of a focal softening induced by the closure of an artery whose walls have been the seat of a syphilitic endarteritis. It develops slowly, and is often preceded by prodromal headaches and vertigo; often there develop other focal lesions, manifesting themselves at times in ocular palsies, paralysis of muscles on the other side of the body, disturbance of speech such as anarthria. The proof of an existing or preëxisting syphilis will assist in determining the etiological factor of the hemiplegia. Very often antisyphilitic treatment produces a favorable result.

4. *Simple focal softening due to arterio-sclerosis.* The diagnostic factors in this case are old age, existing arterio-sclerosis, absence of syphilis, no improvement with antisyphilitic treatment.

5. *Toxic causes.* It occurs with uræmia, in the last stages of carcinoma and phthisis, and is then evanescent and atypical in character.

It is wise to regard tentatively that hemiplegias occurring in the young are of syphilitic basis, and to always subject all those patients in whom there is a doubt as to the etiological factor to antisyphilitic treatment.

Localization of the lesion giving rise to hemiplegia may often be determined by the simultaneous involvement of the cerebral nerves and speech.

The following focal lesions should be chiefly borne in mind: 1. Hemiplegia with motor aphasia indicates a lesion of the third left frontal convolution. 2. Hemiplegia with lower facial palsy, a lesion in the posterior half of the internal capsule. 3. Hemiplegia with hemianæsthesia, a lesion in the posterior division of the same half of the internal capsule. 4. Hemiplegia with alternating (crossed) ocular palsy, a lesion in crus cerebri. 5. Hemiplegia with alternating (crossed) facial paralysis (*Gubler*), a lesion of the pons. 6. Hemiplegia with anarthria and deglutitory palsy, a lesion of the medulla oblongata.

PARAPLEGIA

It is necessary to decide whether the paraplegia is the result of a *lesion of the spinal cord* or of a *peripheral neuritis*. All *spastic* paraplegias may be assigned to spinal cord lesions. The following features must be determined in a *flaccid*[1] paraplegia: 1. Whether the sphincters (bladder, rectum) are involved, a condition which arises only in spinal cord disease. 2. The *reflexes*; they are absent in neuritis, and are exaggerated in spinal cord disease. 3. The etiology; paralysis due to alcohol is of peripheral origin.

In a few individual cases it may be extraordinarily difficult to differentiate between poliomyelitis (a disease of the gray substance of the cord) and a peripheral neuritis, because in exceptional cases the involvement of the sphincters occurs in neuritis; just as the reflexes may in the same manner be exaggerated in the hyperæsthetic stage of a neuritis, so may the reflexes be absent in a deep myelitis which destroys the reflex arc.

[1] May we use this adjective to denote what is expressed in the German "schlaffen Paraplegien"? The author means the paraplegia associated with flaccid, limp, or relaxed muscles, as the antithesis to the spastic paraplegia.—THE TRANSLATORS.

Should the diagnosis of a cord lesion be made, it is necessary to establish the following data: —

1. The seat and extent of the disease; paralysis of both legs indicates a lesion in the lumbar and lower dorsal segments of the cord; paralysis of both legs and arms, a lesion of the upper dorsal and cervical segments.

2. The nature of the pathological process, whether a myelitis, a tumor (aneurysm very seldom occurs), carcinoma, tubercular caries, or syphilis. A tumor, or caries, must be palpable (the vertebra must be carefully examined), carcinoma and tuberculosis in other organs must be demonstrated in order to establish a positive result. The clinician's thoughts should turn to aneurysm if there be a concomitant arterio-sclerosis or an idiopathic cardiac dilatation; to syphilis if the cranial nerves be also involved, or if a specific infection is conceded.

In all doubtful cases antisyphilitic treatment should be given.

Paralyses of the Cerebral and Spinal Nerves

Paralysis of the cranial nerves. This is recognized by the defective function of the muscles supplied by them, and can be determined by the anatomical relations (Fig. 17).

The most important signs are here mentioned: —

Olfactory lesion is to be determined by disturbance of the sense of smell, which should be tested by the use of odorous but not irritating substances (musk, asafœtida). Still one must bear in mind the possibility of disease of the nasal mucous membrane and of nasal occlusion.

Optic nerve lesion will be indicated by diminution of vision, limitation of the visual field, color-blindness. (In every case an ophthalmoscopic examination should be made.)

Hemianopsia or *hemiopia* (the loss of sensation in one half of the retina of each side) depends upon lesion of the occipital lobe or of the optic tract as far as the chiasm. The focal lesion is cephalad to the corpora quadrigemina in homonymous hemianopsia when the pupillary light reaction is normal; on the other hand, should the pupillary reflex be absent for those visual rays which fall upon the paralyzed halves of the retina, the cause of the hemianopsia must be sought ventral to the corpora quadrigemina or in the optic tract. The latter condition is called *hemianopic iridoplegia* or *hemiopic pupillary reaction*. *Amblyopia* or *amaurosis* of one eye

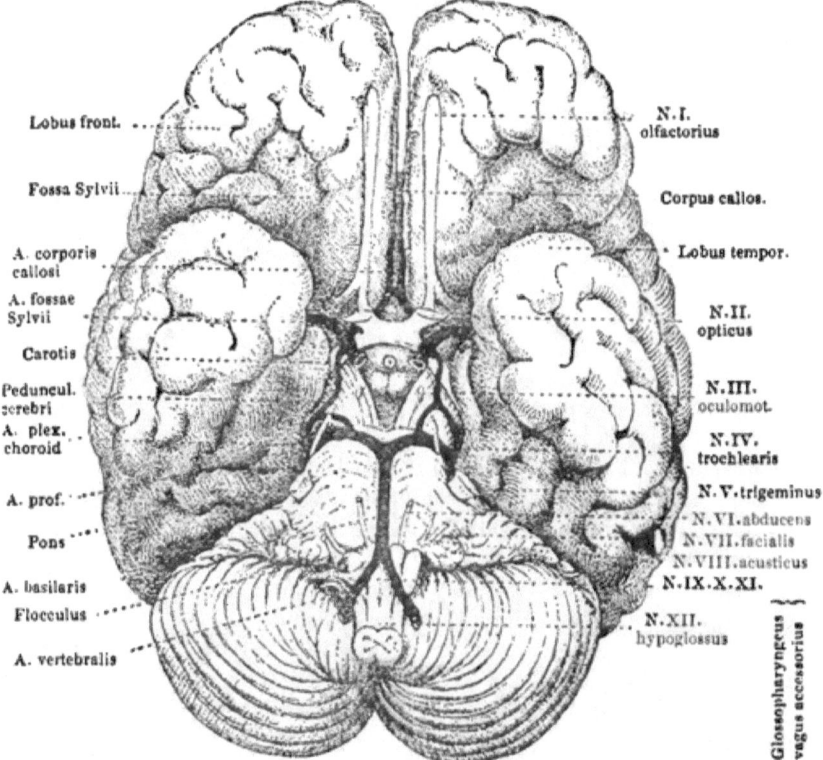

FIG. 17. — THE BASE OF THE BRAIN AND THE CRANIAL NERVES. (AFTER HENKE.)

indicates lesion of the optic nerve anterior to the chiasm; it may also be induced by toxic influences (tobacco amblyopia, uræmic amaurosis).

Motor oculi paralysis is shown by a divergent strabismus, dilatation of the pupil (*mydriasis*), double vision (*diplopia*), closure of the upper eyelid (*ptosis*).

Recurrent ocular motor paralysis accompanies conditions of migraine, which may last for weeks or months (migraine ophthalmique of Charcot), but in such cases the doubt of an organic cerebral lesion (paretic dementia) should always be entertained.

Mydriasis may also be produced by an irritation of the sympathetic, such as results from migraine, trauma, pressure, diseases of the cervical cord, poisoning from atropine, cocaine, etc. It may be also reflex, as in great fright, severe dyspnœa, intense pain.

Myosis, contraction of the pupil, depends upon an irritation of the motor oculi or lesion of the cervical sympathetic; it is unilateral in migraine, and may be the result of the pressure of a tumor on the cervical sympathetic; it is bilateral in affections of the cervical cord and in atropine and pilocarpine poisoning, etc. Concerning reflex iridoplegia, see p. 56.

Trochlearis paralysis is shown by the inability to move the eyes upwards and outwards.

In *abducens* paralysis the eye can not be moved outwards.

Certain abnormal positions of both ocular globes point towards central focal lesions (associated palsies, *nuclear* palsies); for instance, both eyes are immovably fixed looking to the right.

When the muscles of mastication will not perform their function, it indicates a paralysis of the motor division of the *trigeminus*.

Facial palsy is indicated by a loss of function of the muscles of expression. The exact seat of the lesion which produces the interruption of motor conduction may be determined according as the sense of taste, the salivary secretion, hearing, the uvula may be involved (*Erb's* scheme).

Disturbances of hearing may indicate an *acusticus* affection. An otoscopic examination should, however, decide.

Glossopharyngeal paralysis is indicated by a disturbance of taste in the posterior part of the tongue.

Hypoglossal lesion is shown by the deviation of the tongue to one side or the other.

Spinal accessory lesion produces a paralysis of the sterno-cleido-mastoid and trapezius.

Pneumogastric lesion presents an increase in pulse beat and a slowing of respiration.

Lesions of the nerves of the cord are recognized by the loss of function of the muscles which they supply.

Only the symptom groups which commonly appear are here noticed:—

Erb's paralysis is the paralysis of the deltoid, biceps, brachialis anticus, supinator longus, and the infraspinatus. It is due to a lesion of the brachial plexus (5th to 8th cervical and 1st to 2d dorsal nerves).

Median nerve lesion. In this affection, it is impossible to pronate the forearm and flex the hand; likewise the flexors of the thumb, and their antagonists, and the flexors of both the last phalanges are involved. The proximal phalanges can be flexed, and the third, fourth, and small fingers are capable of motion.

Ulnar nerve lesion. Flexion and motion of the hand to the ulnar side and flexion of the third finger are disturbed. The small finger is motionless. The proximal phalanges can not be flexed. Extension of the terminal phalanges of the last four fingers, and abduction and adduction of them are impossible. When the affection is of long standing, the characteristic picture of the "claw hand" is presented; in this case the first phalanges are strongly extended, the end phalanges completely flexed, and the interossei muscles are much atrophied.

Radial nerve lesion. The hand hangs in the condition of flexion, perfectly limp, and can not be extended. The fingers are flexed, the first phalanx can not be extended. The thumb is flexed and adducted and can neither be abducted nor extended. The extended forearm can not be supinated; in the condition of flexion, however, supination may occur by action of the biceps.

Phrenic nerve lesion produces paralysis of the diaphragm, with its characteristic modification of respiratory movements; that is, forcible superior thoracic respiration without the inspiratory protrusion of the epigastrium.

Peroneal nerve lesion produces the characteristic limp drooping of the foot, especially noticeable on walking. Dorsal flexion of the foot and toes, abduction of the foot, and elevation of its outer border are impossible. When of long duration there results a permanent position of the foot which necessitates the walking on the toe tips (*pes equinus*).

Tibial nerve lesion produces a loss of plantar flexion of the foot (the patients can not stand on their toe tips) as well as adduction of the foot and plantar flexion of the toes.

The diagnosis of lesions of the cranial or of the spinal nerves requires, irrespective of a knowledge of the anatomical relations, an investigation of the following data: 1. The *cause* must be determined, whether trauma, pressure, exposure, or an infectious inflammation due to acute or chronic disease. 2. The *intensity* of the disease. This is recognized by the character of the electrical reactions (see p. 64). Three forms are recognized: *a.* The *mild form*, characterized by a normal electrical reaction of the paralyzed muscles. *b. Erb's middle form* by a partial reaction of degeneration. In these cases the irritability of the nerve diminishes without being extinguished, while the galvanic irritability of the muscles on direct application of the current to them is increased; the anode closure contraction is greater than the cathode closure contraction, $AnCc > CaCc$; all contractions are slow. *c.* The *intense form*, characterized by the complete reaction of degeneration, which manifests itself in a complete loss of faradic and galvanic irritability of the nerves, loss of faradic irritability of the muscles, in a quantitative and qualitative change in the galvanic irritability of the muscles; the contractions resulting from the application of an electrical stimulus are slow and wavelike. The anode closure contraction is equal to the cathode closure contraction, $AnCc = CaCc$.

Speech disturbance. — It is necessary to distinguish between speech disturbance due to functional inactivity of the muscles (*anarthria*) and defective speech production where the muscular apparatus is wholly intact (*aphasia*). Anarthria is the result of a lesion of the medulla oblongata (bulbar symptom).

In aphasia one must determine whether the patient has a true conception of the spoken word and is prevented only in transferring the proper concept into speech (*motor aphasia* or *ataxic aphasia*), or whether the conception of speech is so lost that he does not understand the sense of the word pronounced to him and is unable himself to produce word concepts (*sensory aphasia* or *amnesic aphasia*). The site of the lesion producing ataxic aphasia is in the convolution of Broca, the third left frontal convolution; that of sensory aphasia is in the first left temporal convolution.

Motor aphasia very frequently is associated with the inability to write, though the motor apparatus of the upper extremity be intact (*agraphia*), while sensory aphasia may be associated with an inability to read, though perfect vision may be present (*alexia*).

Ataxia is the inability to execute complex movements in a skilful manner when there is no disturbance of the motor apparatus; in other words, the inability to properly coördinate muscular action. It is most probably induced in this wise: after the destruction of the centripetal conduction paths, the movements are no longer controlled by the fine sensory impressions. Ataxia is the chief symptom of *tabes;* it occurs also in peripheral neuritis, is sometimes consecutive to alcoholism and diphtheria; in addition, it occurs in lesions of the cerebellum.

Ataxia of the hands is determined by having the patient, with his eyes closed, button his coat, by having him write, etc. Ataxia of the legs may be demonstrated by having the patient touch the knee of the left leg with the foot of the right, or to describe an arc of a circle, etc. Ataxia is more exaggerated when the patient is in the dark.

Romberg symptom is the swaying of the body while standing with closed eyes. It occurs chiefly in tabes, but may also be present in neurasthenia.

Gait. Slight disturbances of motion or of coördination of the legs may be distinctly recognized in the patient's walk; thus the spastic, the paretic, and the ataxic gait are spoken of.

Motor Irritative Phenomena

Spasm. We designate clonic, that is, interrupted, short contractions, when involving the entire body, as *convulsions;* long-continued contractions are known as *tonic spasms;* when spread over the greatest part of the skeletal muscular system, it is called *tetanus.*

Clonic and *tonic* spasms appear in : —

1. *Epilepsy.* In this affection they are first tonic, subse-

quently they become clonic, and are accompanied by total loss of consciousness, which persists throughout the attack, with dilated pupils, which will not react. The face is at first pale, then becomes cyanosed, and the tongue is often bitten.

2. *Eclampsia* of pregnant and parturient women, commonly accompanied by dropsy and albuminuria.

3. *Uræmia* appearing in the course of acute and chronic nephritis; in a few cases the nephritis may have been unperceived, and the first indication of the uræmic character of the spasms may only have been determined by the discovery of albumin in the urine.

4. As the result of a direct irritation of the cerebral motor centres, occasioned by *tumor, abscess, cysticercus*, etc.

5. In children as a consequence of the *increased reflex irritability* accompanying febrile diseases, dentition, indigestion, worms.

Pure tonic spasms are a feature of: —

1. *Tetany.* They are chiefly confined to the flexor muscles of both arms and both legs. The duration of the attacks may vary from minutes to hours, seldom days. As a rule, there are several attacks daily. The attack may be produced by pressure upon the larger arteries and nerve-trunks of the arm (*Trousseau phenomenon*). The temperature is normal. During the period of repose after the subsidence of an attack, the mechanical and electrical irritability of the peripheral nerves is increased. The *prognosis* is favorable.

2. *Tetanus.* The tonic spasms in this disease are confined chiefly to the muscles of the face, producing *trismus* and *risus sardonicus;* to the muscles of the back, resulting in *opisthotonus*, and to the abdominal muscles; those of the arms and legs are not as often affected. The continuous spasms are interrupted by attacks of jerking. The temperature of the body is increased, and hyperpyrexia is marked before death.

In cases of tetanus the diagnosis should also embrace a determination of the cause and the search for the existing wound through which the tetanus-bacillus entered the system. The wound will often be found to be very slight, or even already healed. The prognosis may be more favorably affected at times when the foreign body (splinter, etc.) which caused the wound has been removed. It is also important to determine the period of incubation; the longer the duration of this, the better the prognosis.

Localized spasm in the course of distribution of certain nerves results partly as a reflex phenomenon, and partly as an individual disease in neuropathic subjects.

Localized tonic spasm occurs along the course of the trigeminus, in the muscles of mastication (trismus), in tetanus, meningitis, epilepsy, hysteria. In such cases artificial feeding may have to be used.

Painful spasm of the legs (cramp) is often noticed after strong muscular effort, in the hysterical and the alcoholist, and in isolated cases of gout and diabetes.

Rhythmical contractions which repeat themselves and which are partly continuous and partly occur at intervals are called *tic convulsif*. We recognize tic convulsif in the distribution of the facial (spasm of the muscles of mimicry), of the spinal accessory (tic of the sterno-cleido-mastoid), seldom, though, in the distribution of the spinal nerves (tic of the rectus abdominis, of the psoas).

Intention spasm or tonic spasm, occurring at the beginning of voluntary movements of the muscles, is the pathognomonic symptom of *Thomsen's* disease (*myotonia congenita*). This disease is one of lifelong duration.

Every voluntary muscle which had been previously at rest becomes, when put into action, the seat of a mild tetanic contraction; the patient can not immediately relax the muscle at command, he is, therefore, unable to carry out coördinated movements; after long tedious motion the contraction becomes less intense. The electrical reaction is peculiarly altered, giving rise to the myotonic reaction of Erb.

Tremor in muscles at rest is the sign in nervous people of intense psychical excitement. *Persistent* tremor without any pathological significance occurs often in old people (*tremor senilis*), but it is characteristic of chronic *alcoholism* and of

Basedow's disease. Extensive tremulous movement is significant of *paralysis agitans*.

The tremor is slow (5 to 6 oscillations in a second) in the senile, in sclerosis and in paralysis agitans; it is quicker (10 to 12 oscillations) in alcoholics and in *Basedow's* disease.

Tremor in voluntary muscles which are put into exercise (*intention tremor*) is pathognomonic for multiple sclerosis; it disappears during sleep, however.

Tremor of the eyes (*nystagmus*) occurs in multiple sclerosis, in workmen who having been confined in the dark are suddenly exposed to light (nystagmus of miners), in hysteria, and in certain nervous affections of the eyes.

Choreic movements, involuntary uncoördinated movements, which embarrass voluntary movements and interrupt them, and which only cease during sleep, are pathognomonic of *chorea* (*St. Vitus's dance*), which is a functional neurosis. They occur very seldom in cerebral lesions.

Choreic movements also appear as an indication of an intense cerebral intoxication in infectious diseases, as in typhoid and miliary tuberculosis.

The following are not of essential significance:—

Athetoid movements. They are involuntary peculiar abductions and flexions of the fingers indicative of a special affection (*athetosis*) or are a symptom of certain central nervous affections (especially of cerebral palsies of children).

Imperative movements are noteworthy in lesions of the cerebellum; they occur as coördinated spasmodic movements (laughing spasms, screaming spasms, jumping spasms) in hysteria and epilepsy.

Cataleptic rigidity of the muscles is noticeable in hysteria, in the hypnotic state, also in meningitis, in certain psychoses (*melancholia attonita*). The limbs remain fixed immovably in any position in which they may be placed.

THE REFLEXES

It is necessary to distinguish between skin and mucous membrane reflex, tendon reflex, and reflex function, whose relations to each other may often be entirely different.

By *skin reflex* is meant the muscular contractions induced in a reflex manner by irritation of the sensory nerves of the skin.

This reflex may be elicited by tickling, pricking, stroking, by cold (touching with ice).

The usual examination comprises a test of : —
1. The *plantar reflex*, produced by irritating the sole of the foot, which is followed by a dorsal flexion of the foot and even of a flexion of the lower extremity on to the abdomen.
2. The *cremaster reflex*. Stroking the inner aspect of the thigh produces reflex movement of the testicle.
3. The *abdominal reflex*, produced by irritating the skin of the abdomen, which induces a contraction of the abdominal muscles of the same side.
4. *Gluteal, scapular, mamillary* reflexes are of slighter importance, and are often absent under normal conditions.

Diminution or *loss* of the skin reflex always results when the reflex path (centripetal nerve, anterior cornu of the cord, and motor nerve) is interrupted; therefore it is to be found in disease of the peripheral nerves and of the spinal cord.

Increase of skin reflex occurs in : 1. Cases of increased irritability of the parts concerned in the reflex arc : hyperæsthesia of the skin, strychnia intoxication, certain neuroses. 2. Cases of restoration of certain interrupted processes, as in diseases of the brain and spinal cord.

The mucous membrane reflexes are : 1. *Conjunctival* and *corneal* reflexes, which manifest themselves in closure of the eyes when the conjunctiva and cornea are respectively touched. 2. *Retching*, by irritation of the pharynx. 3. *Sneezing*, when the nasal mucous membrane is irritated. 4. *Coughing*, on irritation of that of the larynx and respiratory passages.

Tendon reflex is the term applied to the contraction of a muscle which is induced by irritation of its tendon, periosteum, or fascia.

1. *Patellar reflex* (knee jerk, knee phenomenon). If the patellar tendon be struck by a percussion hammer [1] while the leg is flexed and hangs perfectly flaccid or limp, the leg is thrown into a forwards and backwards movement by the resulting contraction of the quadriceps extensor muscle. It is necessary in eliciting this phenomenon that the attention of the patient be withdrawn from his knee. For this purpose it is well to make use of *Jendrassik's* trick: to have the patient clasp his hands across his chest and to pull at them with all his strength; at the moment in which he is exerting the pull, the tendon should be struck unexpectedly.

2. *Achilles tendon reflex*. The Achilles tendon should be suddenly struck while the foot of the patient is slightly flexed; the gastrocnemius will thereupon distinctly contract. This reflex may be absent even in the normal person.

3. *Foot clonus* (foot phenomenon). The foot should be suddenly and forcibly flexed dorsally while the knee is slightly flexed; whereupon there follows an energetic tremor of the foot. It is very seldom found in health.

4. Tendon reflexes of the upper extremity appear very seldom in normal conditions.

The *loss* of tendon reflexes occurs whenever there is an interruption or break in the reflex arc. The reflex path is by way of the sensory nerves, the posterior columns of the cord, the anterior cornua, and the motor nerves. Hence the tendon reflexes are absent in all peripheral lesions, such as multiple neuritis, diphtheritic and alcoholic neuritis, and traumatic lesion, in the degenerations of the posterior columns, as in tabes, and in the diseases of the gray substance of the lumbar cord, as in poliomyelitis.

Recently the loss of tendon reflex has been demonstrated in cachectic conditions as they occur in pernicious anæmia, and in the grave form of diabetes.

Exaggeration of reflexes occurs whenever the reflex inhibitory centres are diseased. These centres are probably situated in the brain, and send out their impulses along paths which proceed through the pyramidal tracts of the

---

[1] The finger tips will answer the same purpose. — THE TRANSLATORS.

cord. Therefore exaggeration of the reflexes will be found in cerebral paralysis, as well as in chronic myelitis; that is, in that form which gives rise to spastic paralysis.

Reflex functions: 1. Pupillary reaction to light and to accommodation.

The pupillary sphincter muscle is supplied by the motor oculi nerve and the pupillary dilator by the sympathetic. Irritation of the motor oculi results in contraction of the pupil; irritation of the sympathetic produces pupillary dilatation. To these conditions the respective names of myosis and mydriasis are given. Paralysis of the motor oculi produces dilatation, that of the sympathetic, contraction of the pupil. The centre is said to be placed in the inferior cervical segment of the cord (*centrum cilio-spinale*). (Compare p. 47.)

The pupillary reaction may be absent in various cerebral affections. *Reflex iridoplegia* is of very great importance as a diagnostic sign of tabes dorsalis. It shows itself as follows: the pupil contracts on accommodation, but not to light. At the same time it very frequently occurs in tabes that the pupils are extremely small (*myosis spinalis*) and unequal in size. Reflex iridoplegia further is an early symptom of paretic dementia.

2. Disturbances of defecation and of micturition, as well as of the sexual reflex (vesical tenesmus, incontinence, constipation, but seldom incontinence of fæces) are pathognomonic of disease of the lumbar cord, especially of tabes and diffuse myelitis. Impotence often appears early in locomotor ataxia.

Impotence is also a symptom of diabetes and Bright's disease, as well as of conditions of irritable weakness or exhaustion in neurasthenics.

Of less diagnostic importance is the direct mechanical irritability of the muscles, which, as a rule, are well preserved.

Paradoxical contraction of Westphal, which occurs among other diseases in multiple sclerosis, paralysis agitans, has been up to this time of little value. It consists in the fact that when the foot is passively flexed dorsally, it will remain for some minutes in this position, even after being left alone, and the tendon of the tibialis anticus muscle becomes prominently visible.

The Disturbances of Sensation

The diagnosis of every disease of the nervous system can only be completed after a thorough test of sensation. Disturbances of sensation are pathognomonic of many diseases of the brain and spinal cord; for instance, lesion of the internal capsule, tabes, syringomyelia, neuritis. In many other cases the extent and intensity of the diseased process can only be determined by an examination of the sensation.

We designate as *anæsthesia* the loss or diminution of sensation; *hyperæsthesia* the increase of the same so that very weak irritations are perceived as strong ones. *Paræsthesias* are abnormal sensory perceptions, such as itching, crawling, formication, the sensation of being stroked with fur, etc. *Neuralgias* are pains, chiefly paroxysmal, in the course of certain nerves. *Pressure points* are those points where the nerve appears directly beneath the skin or on the bone which are especially painful on pressure.

The neuralgias are special affections, as, for instance, trigeminus neuralgia, supraorbital neuralgia, sciatica.

There are special forms of neuralgia which recur paroxysmally at a definite period of the day and which are occasioned by malarial infection (malarial neuralgia, latent intermittent); these forms succumb to large doses of quinine.

A complete test of sensation must regard the various *qualities of sensation* of which each may be tested in various ways (partial anæthesia).

We should test tactile sense, pain sense, space sense, pressure sense, dynamic sense, muscular and articulation sense, temperature sense, and electro-cutaneous sense.

These tests should be made at various periods, as the undivided attention of the patient is necessary.

1. *Tactile sense.* The various regions of the skin should be touched lightly with a brush, the eyes of the patient should be closed, and he should be directed to say "yes" at each touch. In some cases, especially in tabes, the answer of the patient may come some time after the touch (delayed conduction).

Hyperæsthesia in the extremities occurs among other symptoms as a sign of neuritis; it occurs also in the prodromal stage of some acute infectious diseases, as in typhoid, for instance, and during the entire course of meningitis.

2. *Pain sense* is tested by pricking with a pin, by pinching, and with strong electric currents. *Analgesia* is the lack of sensation (irrespective of the ordinary sense of touch) to the usual painful irritants, such as the prick of a pin; it occurs in hysteria, tabes, and peripheral neuritis. In some individual cases a *delayed painful sensation* may be demonstrated; at times tactile sensation and pain sensation are felt one after the other (*double sensation*).

3. *Space sense or sense of locality.* The patient should be touched and he should immediately indicate the spot which was touched or place his finger upon it. The examination of the *tactile distance* belongs under this head. It is determined by the repeated application of a compass with widely opened legs, which are brought closer and closer together until the smallest distance apart in which they are felt as two tactile impressions is registered. This distance in the normal individual is for the

Tip of tongue, 1 mm.
Tip of finger, 2 mm.
Mucous surface of lip, 3 mm.
Dorsal surface of 1st and 2d phalanges and inner surface of finger, 6 mm.
Tip of nose, 7 mm.
Thenar and hypothenar eminences, 8 mm.
Chin, 9 mm.
Tip of large toe, cheeks, and eyelids, 12 mm.
Heel, 22 mm.
Dorsum of hand, 30 mm.
Throat, 35 mm.
Forearm, leg, and dorsum of foot, 40 mm.
Back, 60 to 80 mm.
Arm and thigh, 80 mm.

4. *Pressure sense.* Considerable disturbances of this sense are recognized by applying various strong degrees of pressure with the

finger, slighter disturbances by applying various heavy weights upon the extremities, which ought to be supported upon a firm base. The patient should mention the differences between the weights. A healthy person can recognize differences of one-tenth of the original weight; the smallest difference appreciable varies from 0.005 to 0.5 g.

5. *Dynamic sense* is tested by lifting in the hand various heavy weights which have been wrapped in a cloth and estimating their weights. In the normal person the dynamic sense is finer than the pressure sense.

6. *Articulation sense and muscle sense.* By means of these senses the position of each limb is determined with closed eyes. It is tested, for example, by flexing the fingers of the patient to smaller and larger degrees and fixing them, when the patient is to recognize the various degrees of excursions or flexions with closed eyes; or the leg is to be abducted and fixed in the position in which the slightest degree of abduction may be recognized by him. Disturbances of the articulation sense may be demonstrated in all cases where there is an interruption of the sensory conduction tracts, particularly in tabes and neuritis, mostly before the gross ataxic disturbances become distinct.

7. *Temperature sense.* The skin is touched with dry test tubes or metal cylinders, some of which have been filled with cold and some with warm water. In health, differences of $\frac{1}{2}°$ to $1°$ C. may be detected between $25°$ and $35°$ C. marks of the thermometer. *Perverse temperature sense* is the term given to the disturbance of sense in the patient who recognizes a sensation of heat when touched with ice. Another more recent test of the temperature sense consists in comparing, in the various regions of the skin, the absolute temperature sensibility which under normal conditions show constant fixed local differences (*Goldscheider*).

8. *Electro-cutaneous sensibility.* A faradic current is applied with a wire brush and the figure is read from the instrument after the secondary coil is pulled out to the extent at which the current is first felt.

Examination of Electrical Irritability

The examination of the electrical irritability of paralyzed muscles and nerves leads to important conclusions concerning the location of the disease, but especially to a more definite prognosis of the duration of the disease.

Method and normal relations. — An electrical examination is conducted with the faradic (induction) and with the galvanic (constant) current. The indifferent electrode is placed upon the sternum, the other upon the muscle or nerve to be examined. The excitation of the muscle by way of the nerve is called *indirect irritation*; that resulting from the direct application of the electrode to the muscle itself, *direct irritation*. The examining electrode for determining the irritability of each muscle or nerve is placed upon certain fixed points which have been determined empirically and may be seen in Figs. 18 to 22.

When the faradic current is applied by direct and indirect irritation, a distinct contraction results. The distance in millimetres to which the secondary coil is moved to produce the slightest noticeable contraction is then read. It is denoted in all German books as R.A.—mm.

In the *galvanic examination* the electrode used is converted by means of a *pole changer* to first the *negative* (*cathode, zinc*) pole, then to the *positive* (*anode, carbon,* or *copper*) pole. On gradual increase of the strength of the current the first weak contraction of the muscle appears, providing the current is closed in such a manner that the examining electrode is made the cathode pole (CaCc). The amount of current used (determined by the number of cells used or by the absolute galvanometer) is then registered. With this amount of current, reaction will not yet appear on cathodal opening, anodal closure, or anodal opening. By increasing the strength of the current a gradual appearance of contraction will appear on anodal opening (AnOc) and on anodal closure (AnCc); only with a much stronger current than this will cathodal opening contraction appear, and then the cathodal closure contraction will already have been tetanic. The order in which contractions of a *normal* muscle takes place by increasing strength of current and *indirect* irritation is, hence, as follows:[1] 1. CaCC; 2. AnOC; 3. AnCC; 4. CaCTe; 5. CaOC. Contractions are short and lightning-like.

[1] The following abbreviations are in use: Ca = cathode; An = anode; C = closure; O = opening; c = slight contraction; C = strong contraction; Te = tetanus.

Quantitative change of electrical irritability. In different diseases the electrical irritability of the nerves and muscles is simply increased or diminished without there being any change in the order or in the quality of the contractions.

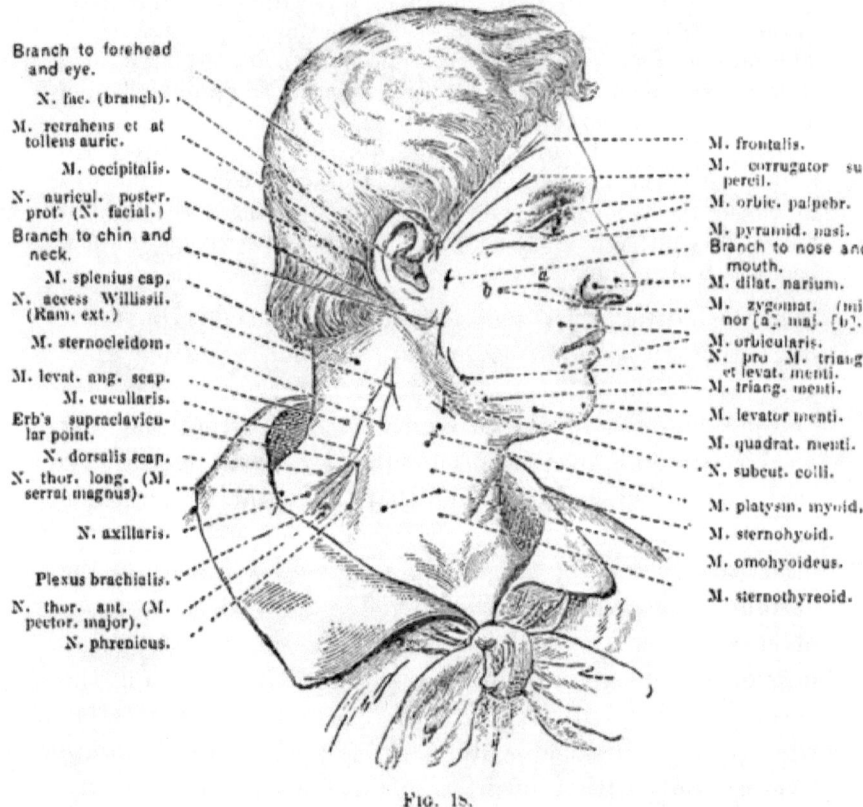

Fig. 18.

In order to compare the results of examination of a unilateral paralysis, the corresponding nerves and muscles of the unaffected side should be tested as well. In bilateral paralysis and in general disease of the nerves a comparison is made with the electrical irritability of certain nerves as it occurs in the normal person, the nerves used being the frontal, accessory, ulnar, peroneal nerves.

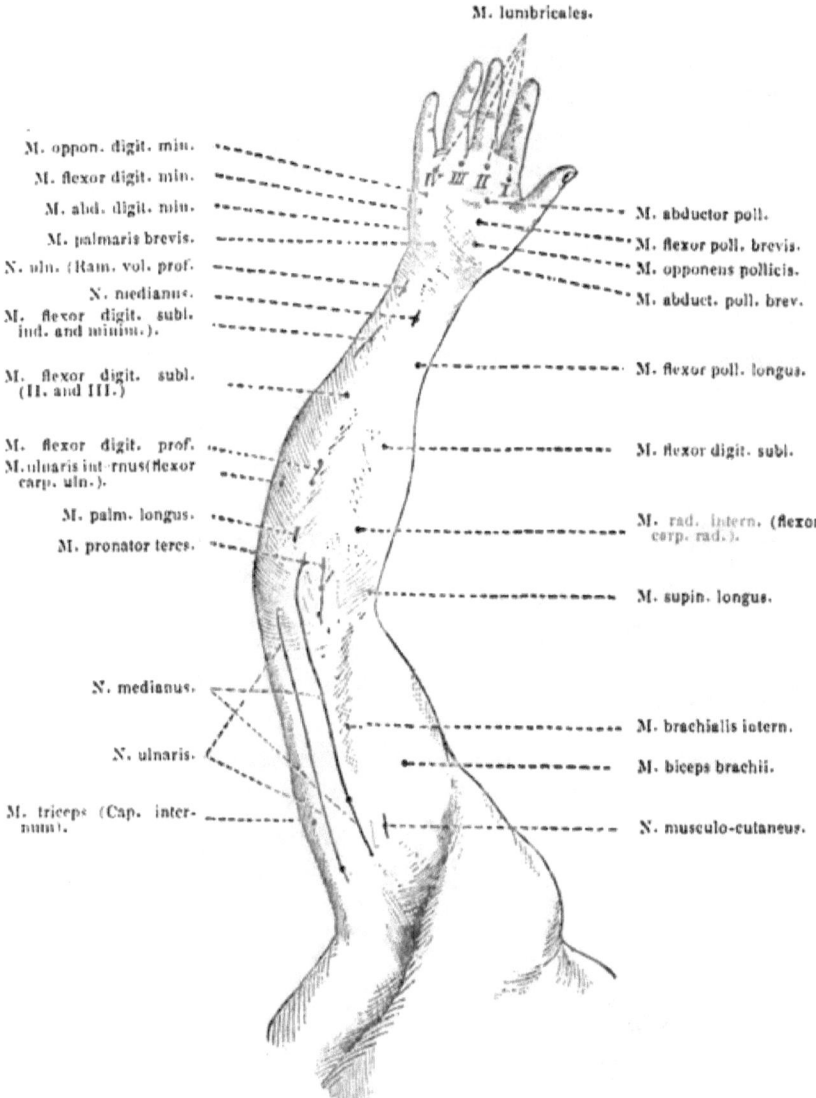

Fig. 19.

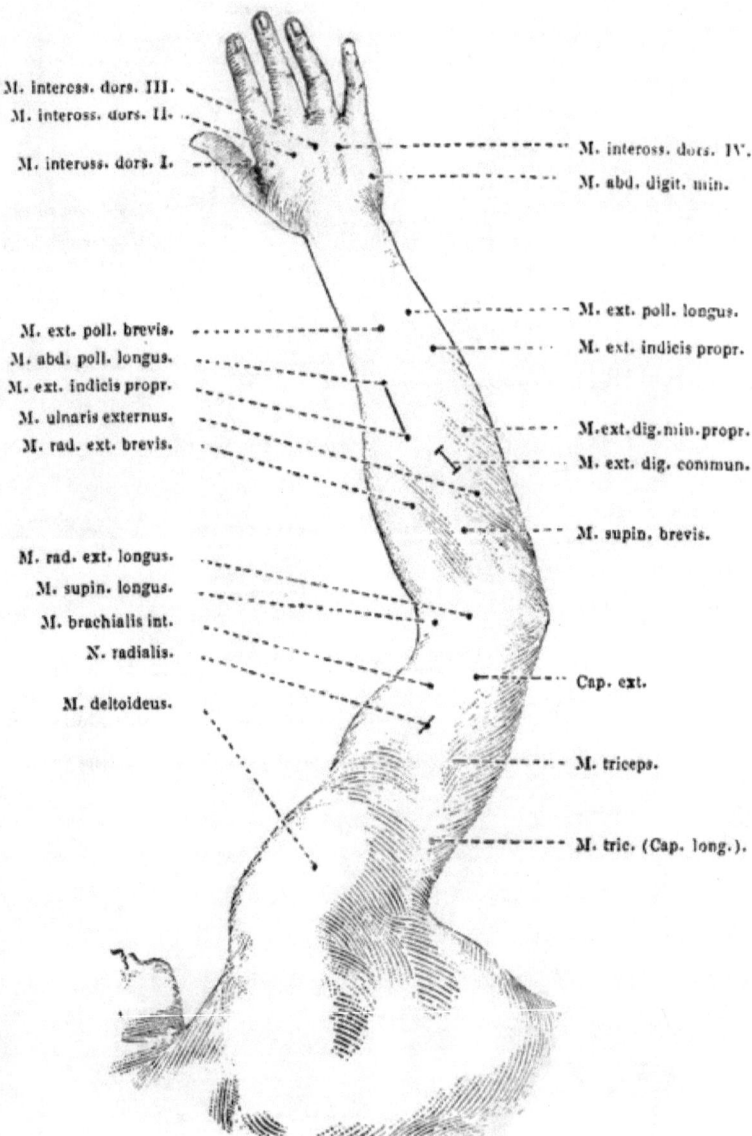

Fig. 20.

Electrical irritability is *simply increased* especially in tetany, and in cases of recent peripheral lesions.

Electrical irritability is *simply diminished* in all paralysis of long standing which leads to muscular atrophy, as in cerebral hemorrhage, bulbar lesion, cord lesion where the trophic centres are intact.

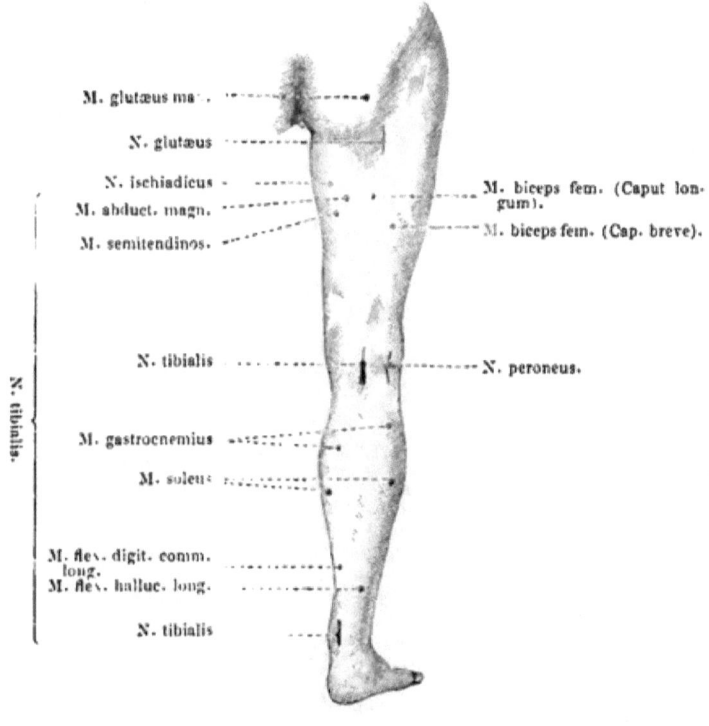

FIG. 21.

Qualitative change of the electrical irritability: Degeneration reaction (R. D.). Qualitative change of the electrical irritability is an infallible sign of paralysis of the nerve *peripherally;* that is, in cranial nerves the gray nuclei, in spinal nerves the ganglia of the anterior cornua are themselves diseased or the disease lies peripherally to these trophic centres. In these cases the *faradic and galvanic irritability of the nerves* diminish more and more a short time after the beginning of the paralysis. In 8 to 10 days the irritability is absolutely lost; no contractions are to be obtained even with strong currents. The direct faradic irritability of the muscles is also abolished.

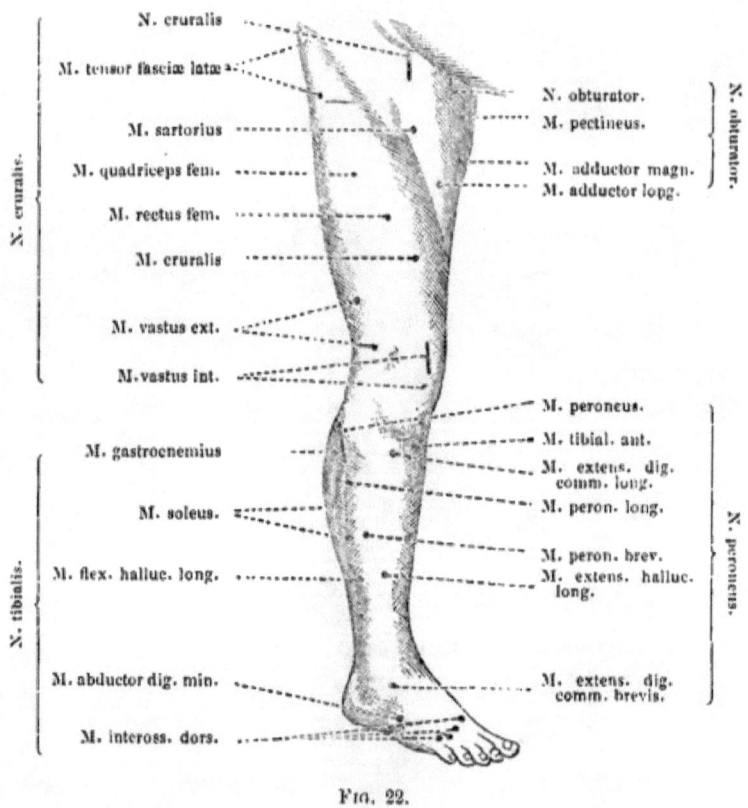

FIG. 22.

On the other hand, the direct *galvanic irritability of the muscles* is appreciably increased from the second week; contractions result even with the weakest currents. The contractions are *slow, long continued,* and vermicular. The AnCC follows upon the CaCC with the same strength of current, and is often stronger than the CaCC; frequently

CaOC appears sooner than AnOC with less strong currents (complete degeneration reaction).

In severe, more especially in incurable, cases the galvanic muscular irritability diminishes after 4 to 8 weeks' duration of the disease, until it completely disappears. In curable cases the conditions of electrical irritability become gradually normal; in such cases the power of voluntary movement of the muscle recovers, however, often much sooner than the electrical irritability.

Partial degeneration reaction, which permits of a diagnosis of paralysis of lighter intensity, consists in this, that faradic and galvanic irritability of the nerves and faradic irritability of the muscles are only slightly lowered, whereas the distinguishing characteristic changes on direct galvanic irritation of the muscle are entirely equal to each other (see p. 49).

The trophic condition of the paralyzed muscles is in harmony with the electrical relation; disease of the ganglia of the anterior cornua, as well as of the nerves peripheral to these, leads to *degenerative atrophy;* disease central to the trophic centres leads only gradually to slight diminution in bulk of the paralyzed muscles.

Degeneration reaction occurs in peripheral lesions of the motor nerves (traumatic and rheumatic, diphtheritic, toxic paralysis, multiple neuritis, alcoholic neuritis), in disease of the gray anterior cornua, and the gray nuclei of the medulla (atrophic palsies of children, poliomyelitis).

Degeneration reaction is absent in all cerebral and in all spinal paralyses whose origin is cephalad to the trophic centres.

Degeneration reaction is of prognostic importance in this, that either irreparable atrophy of the affected muscles will ensue or that recovery can only at best begin to take place in 2 or 3 months. Absence of degeneration reaction is a certain diagnostic sign that no gross anatomical changes have taken place; it is of prognostic value in that it denotes that recovery will ensue in a short time, often in 3 to 4 weeks.

Partial degeneration reaction indicates that the muscles have suffered serious anatomical changes, whereas the nerves have been unaffected; it admits, therefore, as far as time is concerned, of a better prognosis than complete degeneration reaction.

Symptoms of Some of the Diseases of the Nervous System

Cerebral palsies have been spoken of sufficiently in the beginning of the chapter (see p. 43 *et seq.*).

Cerebral abscess is diagnosed by: (1) the etiology (trauma, otitis, embolism); (2) the general symptoms (headache, vertigo, vomiting), varying in intensity; (3) the irregular remissions of the fever; (4) the absence of choked disc; (5) the focal symptoms, which vary according to the site of the abscess, although they may be entirely absent.

Tubercular meningitis is diagnosed when: (1) other organs are demonstrated to be the seat of a tuberculous process; (2) the fever is of irregular remittent type; (3) the pulse is slow and irregular; (4) the stupor is severe and associated with delirium; (5) the neck is rigid, there is vomiting, and the abdomen is retracted; (6) finally tubercle bacilli in the cerebro-spinal fluid obtained by lumbar puncture are found.

Cerebral syphilis may be diagnosed from the following, if proof of previous syphilis can be obtained: there are often intense prodromal headaches, epileptiform convulsions, and apoplectiform attacks, various paralyses. The general signs and the focal symptoms are not at all characteristic, for the same may be present in tumor, hemorrhage, etc. *In all doubtful cases mercurial innunctions should be given*, which, if remedially effective, would determine the diagnosis.

Tumors of the brain. — A diagnosis may be made from the following data: (1) The general symptoms, such as headache, vertigo, vomiting, convulsions, psychic weakness, are ushered in gradually, and increase slowly in intensity; (2) ophthalmoscopic examination reveals choked disc, though an absence of this condition does not disprove the presence of a tumor; (3) there are focal symptoms which vary, of course, according to the region of the brain, which is the seat of the growth.

Progressive bulbar paralysis is accompanied by disturbance of speech (anarthria), atrophy of the tongue, of the lips, and of the muscles of mimicry supplied by the facial nerve. Owing to this paralysis, the face appears mask-like. There is difficulty in swallowing, arising from the atrophy of the pharyngeal muscles. Speech is weak, and its tone unvarying (monotone), owing to atrophy of the laryngeal muscles; coughing becomes impossible, and finally respiratory paralysis supervenes.

Degeneration of the combined pyramidal tracts (manifested by muscular atrophy and increased tendon reflexes in the upper and spastic paralysis in the lower extremities) is associated with degeneration of the nuclei of the medulla. This condition has been recognized by a few authors as a distinct and special disease, which they call *amyotrophic lateral sclerosis*.

In some cases bulbar symptoms are associated with *progressive spinal muscular atrophy* (see below), which consists of a degeneration of the anterior cornua and the anterior nerve-roots.

All three diseases present a degeneration of the same motor conducting apparatus, but in different locations, so that a diagnostic separation in many cases appears artificial.

Myelitis. — The symptoms vary according to the position of the process in the cord.

Myelitis cervicalis is characterized by paraplegia of the legs, paralysis and sensory disturbances in both arms. The tendon reflexes are increased, and spastic phenomena are present. The bladder and rectum are affected.

Myelitis dorsalis presents essentially the same symptoms, but the arms are wholly unaffected.

Myelitis lumbalis. Paraplegia of both legs, but the upper extremities are unaffected. The bladder and the rectum are involved; the skin and tendon reflexes are diminished or entirely absent.

As a special form of chronic myelitis, there have been described: (1) the so-called *spastic spinal paralysis* (*Erb, Charcot*), characterized by spastic symptoms in the extremities, increased reflexes with the preservation of the bladder and sexual functions, and said to depend wholly on lesion of the pyramidal tracts; and (2) *progressive spinal muscular atrophy*, which gradually leads to degenerative atrophy of the muscles of the arm and shoulder, and which is due to a lesion of the gray anterior cornua and anterior roots.

The description of *poliomyelitis* (lesion of the gray substance of the cord) belongs to this category. It is often confounded with multiple neuritis, and its diagnosis is made with great difficulty. Poliomyelitis can only be diagnosed with certainty when it occurs in children, where it is known as the *essential (spinal) palsies of children*. Here the age, the acute onset, the subsequent flaccid paralysis, accompanied with atrophy and the presence of degeneration reaction, the absence of reflexes, and the preservation of sensation, will establish the diagnosis.

Tabes dorsalis is diagnosed by the iridoplegia, the loss of tendon reflexes, lightning-like pains in the legs, the ataxia, the Romberg symptom, analgesia, at times anæsthesia, delayed conduction, and paræsthesia. In some cases there are periodical disturbances of the abdominal viscera, as, for instance, frequent painful attacks of vomiting (*crises gastriques*). The presence of two characteristic symptoms (for instance, iridoplegia and failure of the reflexes, or iridoplegia and *crises gastriques*) will render the diagnosis certain. It is distinguished by three stages: the neuralgic, ataxic, and paraplegic.

Multiple sclerosis of the brain and spinal cord. — In typical cases there is intention tremor, scanning speech, nystagmus, spastic paretic gait, increased tendon reflexes, foot clonus, and gradually developing psychical weakness. Only seldom do we find all signs present in a single case.

CHAPTER IV

DIAGNOSIS OF THE DISEASES OF THE DIGESTIVE SYSTEM

Lips. — The condition of the nutrition (see p. 4) of the patient and the state of the blood (see p. 7) are determined by the color of the lips. Dryness of the lips, and crusts thereon, indicate a febrile disease. In typhoid, the brown, rusty color of the lips (*fuligo*) is characteristic.

Teeth. — A healthy person has moist teeth. Dry teeth covered with sordes indicate neglect and fever. A good nurse should see that the lips, mouth, and teeth of every comatose patient are kept moist and clean.

A good condition of the teeth insures good mastication. When many teeth are wanting, the food is swallowed imperfectly masticated; very often their absence serves as a guide to the diagnosis of chronic gastritis.

The *age* of children may readily be determined by the number of their teeth, and especially at the time of their appearance. A knowledge of the period of the eruption of the teeth is necessary for the diagnosis of many diseases of children.

The milk teeth are twenty (20) in number; on each side above and below, two (2) incisors, one (1) canine, and two (2) molars. They appear from the 3d month to the 3d year of life, usually in the following order: the lower central incisors from the 3d to the 10th month (average 7th month), the upper central incisors from the 9th to the 16th month, the superior lateral incisors from the 10th to the 16th month, the lower lateral from the 13th to the 17th month. The first four molars from the 16th to the 21st month, the four canines from the 16th to the 25th month, and the four second molars from the 23d to the 36th month, as an average, however, between the 24th and 30th months.

The change to the permanent teeth begins in the 7th year of life and proceeds, as a rule, in the same order as does first dentition. The third molars (wisdom teeth) appear from the 18th to the 30th years of life. The permanent teeth are thirty-two (32) in number: two (2) incisors, one (1) canine, two (2) bicuspids, three (3) molars on each side above and below.

Tongue. — An inspection of the tongue is sanctioned by old custom and should begin the examination of the patient. The tongue of the healthy is bright red and moist, and has no tremor when extruded.

In febrile diseases, before the period when a competent nurse takes charge, the tongue is, as a rule, dry, fissured, and covered with discolored fur. Its appearance in typhoid and in scarlet fever is characteristic; it is covered with a white fur, sometimes brown, and with a red stripe on either side in typhoid; it has the strawberry appearance in scarlet fever (strawberry tongue).

In non-febrile conditions the tongue should be examined to see if it be bright red or coated.

A normal condition of the tongue in these respects would in most cases exclude disease of the stomach. A coated tongue often indicates a catarrhal condition of the gastric mucous membrane. Still these conditions are not regularly united, so that the diagnostic deductions from the coated tongue must be drawn with care. It should be borne in mind that ulcer of the stomach (*ulcus ventriculi*) and hydrochloric hyperacidity are usually unaccompanied by a coated tongue.

Glossitis or inflammation of the tongue, attended by swelling of and intense pain in this organ, is an infrequent but severe infectious disease which requires chiefly surgical treatment.

Mouth. — Inflammation of the mucous membrane of the mouth (*stomatitis*) is recognized by swelling and œdema of the mucous membrane which is exceedingly painful. It is occasioned chiefly from neglect and often from the use of mercurials. It is relieved by rinsing the mouth with potassium chlorate.

A mild form of stomatitis not infrequently develops after a generous consumption of grapes such as is ordered in "a grape cure."

A reddish border at the junction of the gums with the teeth occurs often and has no diagnostic significance; a grayish or bluish border is produced by plumbic sulphide and indicates plumbism (lead intoxication); it is the so-called blue line of the gums.

Soor or thrush consists of small grayish-white membranous exudations upon the buccal mucous membrane which occur in weakly children often after neglect and uncleanliness, and develop in the adult chiefly toward the end of some severe disease (phthisis, sepsis, etc.). (See Chap. XII.) The cause of this disease is a fungus (*Saccharomyces* or *Oidium albicans*) which grows in acid media in buds and in alkaline media in threads. In the mouth it grows in threads and round conidia.

Pharynx and tonsils. — (See Chap. V., Diagnosis of the diseases of the upper air-passages.)

Saliva is an alkaline glandular secretion containing mucin, whose active principle is *diastase*, which changes starch into sugar. In stomatitis and in a few diseases, such as diabetes, the saliva becomes acid. In hyperacidity of the stomach the diastatic activity is often reduced. Still, thus far, the examination of the saliva has not attained any diagnostic importance.

Saliva is recognized by the blood-red color obtained by adding an acid and precipitating with ferric chloride. This reaction depends upon the presence in saliva of rhodanpotassium (CNSK).

ŒSOPHAGUS

Of the diseases of the œsophagus *stricture* of this organ has a special diagnostic importance. It is determined by the complaint of the patient that the bolus of food sticks in his throat or before it reaches his stomach, and that it is often regurgitated and tastes very badly.

About the site of the stricture a dilatation is produced by the pressure of the accumulated food, in which decomposition and fermentation of the food may take place. The retained remnants of food are regurgitated after a while by the retching produced from the irritation. We can recognize that these remnants never reached the stomach by the fact that they are alkaline in reaction, owing to the admixture of saliva, and where milk has been taken by the fact that it is not coagulated.

A special form of constriction is the *diverticular* (a dilatation of the mucous membrane *without a primary stenosis*). We distinguish between a *pressure* diverticulum and a *traction* diverticulum. The former is always situated in the extra thoracic part of the œsophagus, reaches a considerable size, varying between 5 and 10 cm., so that when distended it may be seen and felt on the outside of the neck as a distinct *tumor*. This form of diverticulum may shut off the way for the entrance of nutriment and thus cause death from inanition. It is produced by pressure of foreign bodies, swallowing too large masses of food, and by traumatism. The *traction diverticula* occur in the lower half of the œsophagus, and have a minimal size of 4 to 8 mm. They are scarcely ever the object of medical diagnosis. They are produced by the traction resulting from cicatricial formation; in rare cases they have led to pulmonary gangrene and death by perforating into the bronchi.

The diagnosis of œsophageal stenosis is made with certainty by introducing a moderately hard œsophageal sound of ordinary calibre. Before this is done it is absolutely necessary to examine in each case whether there is an aortic aneurysm present; should such be the fact, no attempts at introducing the instrument should be made. Recently success has been attained in obtaining a view of the interior of the œsophageal tube by means of direct œsophagoscopy (*Rosenheim*); but this method of examination is too difficult to admit of general use.

It is necessary to diagnose: 1. *The seat of the constriction*, as the surgical possibility of relief will be thereby determined. To determine this point, mark the place on the œsophageal sound where the teeth touch it after the constriction has been reached, and after the instrument has been removed measure the length of the part which has been included between these two points.

The distance between the upper incisor teeth and the cardia of the stomach measures in adults, as a rule, 40 cm., between the incisors and the beginning of the œsophagus the distance is 15 cm., and from the incisors to the point where the œsophagus crosses the bronchus is 23 cm.

2. *The cause of the constriction*, as from this directly depends the prognosis and treatment. The most frequent cause of constriction is *carcinoma*. This is to be diagnosed in old people, especially if they emaciate rapidly and have a cachectic appearance, if there be no special reasons against it.

In the young the following causes may occasion the stenosis: 1. *Cicatricial formation*, (*a*) the result of an extensive burn by acids or alkalies, which may have been swallowed accidentally or with suicidal intent, (*b*) the result of an ulcer of the cardiac orifice of the stomach or of the œsophagus, which has given symptoms for years of a round ulcer. 2. *Œsophagitis*, which arises from swallowing a foreign body or from the extension of inflammation from a neighboring pus formation. 3. *Tumor of the mediastinum or aortic aneurysm*, which may be determined by the physical signs. 4. *Cheesy bronchial glands* at the hilus of the lungs, which may be presumed to exist when tuberculosis of the lungs accompanies the stenosis. 5. *Syphilis*, which seldom leads to œsophageal stenosis, and which can only be assumed when syphilis is shown to be present and when every other etiological factor has been excluded. 6. *Hysteria* in the young should cause one to think of the possibility of a nervous spasm of the cardia.

The diagnosis is only completed when the **permeability** and the *calibre* of the stricture are determined; one should try to pass the stenosis with carefully repeated applications of sounds of varying calibre. The proof of the **permeability** of the stricture is not always to be found in the descent of the sound, for that may be occasioned by the sound bending upon itself in a diverticulum, but only upon hearing the *deglutitory murmur* when the patient is asked to swallow.

This sound is heard on auscultation: 1. Posteriorly, on the left side next to the vertebral column at the level of the 6th dorsal vertebra; it is a short, dull murmur, which may be heard immediately after the act of swallowing. In stenosis it is exceedingly weak, and in closure of the œsophagus it is entirely absent. 2. Anteriorly, over the arch of the ribs on the left side next to the xiphoid process. Often, in addition to the deglutitory (primary) murmur, another (secondary) murmur, occurring from *3 to 5* seconds later, may be heard, or even a third; these secondary sounds are most probably produced by regurgitating bubbles of air. In stenosis these secondary murmurs are heard from *5 to 12* seconds later.

Above a stenosis and over diverticula, loud sounds, at times lasting many minutes, may be heard on auscultation; they arise from the motion transmitted by muscular action to the fluids retained in the diverticula.

The Diagnosis of the Diseases of the Stomach

As far as the history of the cases is concerned, the following points are to be regarded:—

Hereditary influences have seldom an important bearing, except for carcinoma and neurasthenia. The patient's mode of life, however, is of greatest importance, whether he has been exposed to injurious occupation such as sedentary work, troubles and worries, psychical emotions, lead intoxication, etc., whether he was guilty of *excesses in diet*, or bolting of food, or of chewing his food imperfectly, or of taking his food too hot, whether he was an alcoholist or indulged excessively in tobacco. An essential point to be elicited is whether he became emaciated rapidly.

The next step is to investigate accurately the complaints of the patient concerning the signs of dyspepsia, to leave nothing omitted, and to follow with an inquiry concerning any casual former troubles in the other organs (lungs, heart, and kidneys).

The complaints of all patients with gastric affections, and those by which the attention of the physician is directed to the stomach, refer to the general so-called dyspeptic symptoms, viz. loss of appetite, eructations, heart-burn, pressure and fulness in the region of the stomach, pain in that organ, depression. None of these symptoms in itself is characteristic enough to justify a definite diagnosis of the disease of the stomach. It is of special diagnostic value that the dyspeptic symptoms occur also in the course of diseases of other organs, as, for instance, in the inception and progress of pulmonary consumption (phthisical dyspepsia), in heart diseases during the stage of compensatory disturbance, in nephritis, in attacks of gout, in diabetes.

The appetite is an important indication of health. *Loss of appetite* is a sign of a general pathological disturbance, though no special deductions may be drawn from it. Most of the febrile and chronic diseases are accompanied by it. The special forms of gastric diseases which begin with a loss of appetite are gastritis and carcinoma. Gastric ulcer and hyperacidity, on the other hand, are usually associated with a good appetite. Still numerous exceptions occur. Excessively increased appetite (*bulimia*) as well as its perversion, pathological longings, are chiefly the signs of gastric neuroses, although they sometimes occur in other affections.

It was the custom to regard the condition of the appetite as an infallible sign of either good or bad digestive power; this is, however, only conditionally true. In many cases even a poor appetite may be associated with good digestive power.

Eructation. — This symptom indicates the presence of fermentation in the stomach; it has no differential diagnostic significance.

Frequent loud eructations, spasmodic in character, occur in neurasthenics as a result of swallowing air; the gas discharged is without smell and tasteless; chronic pharyngitis is often present in these cases.

Pyrosis, heart-burn, acid burning in the œsophagus. This is almost always a sign of an increased acidity of the stomach. Still no diagnostic deduction should be made from its presence, because it may just as well indicate an excess of hydrochloric acid (anorganic hyperacidity), a severe fermentation which appears first in cases of deficiency of hydrochloric acid (anorganic anacidity), and which leads to the formation of acetic acid, lactic acid, and butyric acid (anorganic hyperacidity). These conditions of anorganic and organic hyperacidity have different prognostic values, and demand, in fact, different forms of diet and treatment, notwithstanding both may present the same symptom of pyrosis.

Feeling of pressure and fulness in the stomach. This symptom is present in many gastric affections, is frequently purely a neurotic symptom, and is often the result of over-exertion.

Pain in the stomach. This symptom is only to be utilized with the greatest care in making a differential diagnosis. It frequently is present in ulcer, and, on the other hand, in gastric neuroses and catarrhs as well as in carcinoma. Only those pains which are confined to a *circumscribed spot* and which are always felt in that spot are of value in concluding that there is a gastric ulcer. (Localized painful points are found also in the neuroses; see also p. 81.)

Vomiting

Vomiting results when the centre situated in the medulla oblongata is irritated and at the same time the abdominal muscles and the diaphragm are contracted, the pylorus is closed, the cardia is open, and when probably antiperistaltic action is set up in the stomach. The irritation of the vomiting centre is induced either directly by the blood state, as appears when emetics are taken, by swallowing caustic materials such as acids and alkalies, or by poisons such as chloroform and morphia, or by toxic disease-products as in chronic *nephritis, uræmia,* and *cholera,* or by reflex irritation. Reflex vomiting may arise from various organs; from the *brain* in meningitis and tumor cerebri, from the *peritoneum* in peritonitis and appendicitis, from the *kidneys* in renal colic and pyelitis, from the *bladder* in stone, from the *genital organs* in endometritis and *pregnancy,* or from the *stomach* as in several diseases of that organ. The tendency to vomit in *heart disease* is chiefly induced by congestion gastritis, though often acute cardiac dilatation (for instance, induced by over-exertion) as well as cardiac failure are accompanied by it.

Vomiting may also result from psychic influences, especially from disgust. Many persons are nauseated from fright and fear.

Repeated vomiting is a sign of various visceral disorders. It is pathognomonic for meningitis, peritonitis, and uræmia. In Bright's disease it is an evil omen, as it is early indication of ensuing uræmia. Its frequency is for a great part a determining prognostic feature of meningitis. *Pregnancy* is often first diagnosticated by the presence of persistent vomiting, especially when the vomiting occurs with an empty stomach; a special prognostic and very serious symptom is offered by the intractable vomiting of pregnancy, which often may be the indication for the artificial production of premature labor.

Repeated vomiting in *febrile disorders* demands special diagnostic attention. There may be *prodromal vomiting*, as occurs in scarlet fever and erysipelas. It may serve as an *essential symptom of disease*, as in meningitis and peritonitis. It may be caused by *drugs*, e.g., antipyrin, or by *defective nutritive materials*. Its most severe form is associated with "irritative exhaustion" and is then often accompanied by *singultus*, which are chiefly features of the remission of the disease, or of a beginning convalescence.

Periodical recurrent attacks of frequent vomiting which alternate with periods of freedom therefrom have a special diagnostic importance: they are the so-called *gastric crises*, and are associated with distressing, often unbearable nausea. They may last for days and frequently lead to inanition. They appear in diseases of the spinal cord, especially in *tabes dorsalis*. Not infrequently this characteristic group of symptoms may be the first sign which calls attention to the existing tabes which may have previously been overlooked. Still periodical recurrent vomiting may be a sign of a neurasthenic gastric disorder without having a central origin, but in each of the latter cases it is wise to entertain a suspicion of its being of central origin.

From vomiting *alone* no diagnosis of disease of the stomach should be made; there should be a further examination to determine whether any of the following described signs are present.

Vomiting in diseases of the stomach. Repeated vomiting occurs in so many gastric disorders (e.g., ulcer of the stomach, gastritis, dilatation of the stomach, carcinoma, gastric neurosis) that from this sign alone no diagnosis is possible.

Composition of the vomited material. The vomited matter is composed chiefly of particles of food which have been much changed by the gastric juice and by fermentation in the stomach (these changes produce from the carbohydrates, lactic acid, butyric acid, and acetic acid; from the fats, free fatty acids; from the albuminoids, leucin, and tyrosin in addition to the peptones): saliva (especially present in matutinal vomiting); mucus (in gastritis, but not diagnostic); bile (with no diagnostic significance, yet it indicates always that the pylorus is permeable); urea (in uræmia; concerning its demonstration, see Chap. VIII.).

Microscopic examination of the vomited material. Under the microscope we find particles of food (striped muscular fibres, fat globules, starch cells, and vegetable fibres), pavement epithelium, which comes from the mouth and œsophagus, leucocytes, various forms of bacilli and cocci, fungus, and sarcina. An abundance of sarcina indicates a considerable fermentation in a dilated stomach, though it gives no indication as to the cause of the dilatation.

On the other hand, special kinds of vomiting are pathognomonic.

1. *Vomiting of blood (hæmatemesis, melæna)* is pathognomonic of: *a. Ulcer of the stomach,* when the vomited matter is composed of fresh, dark-red, inodorous blood; it also occurs in cirrhosis of the liver. *b.* Vomiting of old, decomposed, sometimes badly smelling blood (coffee-ground vomit) is pathognomonic of carcinoma.

Blood is sometimes vomited when the gastric mucous membrane is inflamed by irritating substances, such as acids and alkalies. In hæmophilia blood may be vomited without an essential cause. Hysterical women in rare cases have hæmatemesis. In young girls vomiting of blood at the time of a suppression of the menstrual flow is of no serious consequence (vicarious hæmatemesis). Still one should always examine carefully for ulcer of the stomach.

One should be on the watch to distinguish between vomiting of blood and coughing of blood. In the majority of cases patients will describe characteristically either hæmatemesis or hæmoptysis. But sometimes coughing is attended by retching, and sometimes blood which has been coughed up is swallowed. In some cases hæmoptysis or hæmatemesis is the first sign of a pulmonary or of a gastric disorder which may have been concealed until the appearance of this sign startles the patient to the highest degree so that he cannot describe accurately the manner in which it appeared. In such cases (seldom occurring) it may be much more difficult to establish a differential diagnosis than in cases of fresh bleeding, where it should be always a rule *to exercise the greatest care in conducting the examination of the organ, or to postpone it until all hæmorrhage has ceased for some time.*

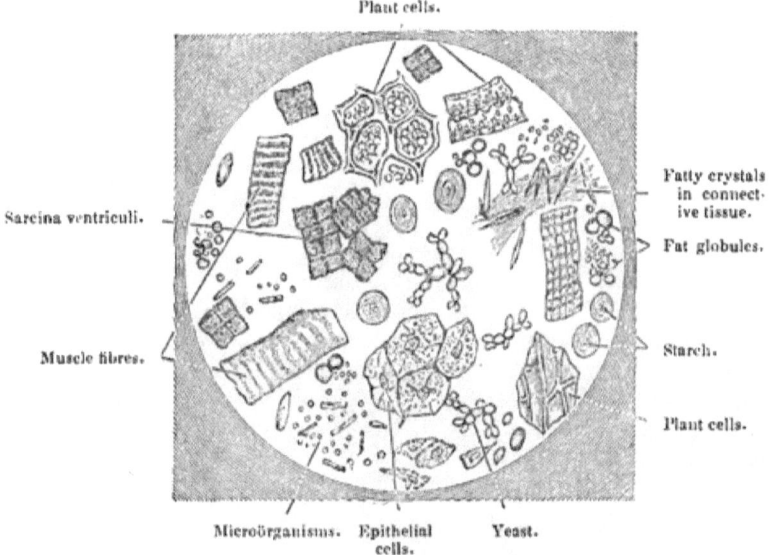

Fig. 23.—Schematic Microscopic Picture of Vomitus.

2. *Fæcal vomiting* (*miserere*) is the sign of intestinal obstruction (*ileus*) (see p. 95.)

3. *Very copious vomiting* appearing after long intervals of time is characteristic of *dilatation of the stomach.*

In a dilated stomach the food accumulates, owing to the atony of the walls or to the inability to pass into the pylorus on account of mechanical obstacles. As soon as the stomach is over-distended by reason of the continuous ingestion of food, it gets rid of a part of its contents by the vomiting of 1 or 2 litres of often decomposed fermented material. After this the patient feels better, eats for several days with steadily increasing discomfort until finally very copious vomiting begins anew.

4. *Matutinal vomiting* or morning vomiting before breakfast, accompanied by intense nausea and composed mostly

of mucus and only seldom of clear fluid, is produced by a pharyngitis such as is most frequently found combined with an alcoholic gastritis (*vomitus matutinus potatorum*).

5. Vomiting *immediately after eating*, chiefly attended by nausea, is characteristic of hysterical or nervous dyspepsia. In such cases one should look always for other signs of neurasthenia.

In most cases it is not possible to make a definite diagnosis from the character of the complaints of the dyspeptics. Under these circumstances it is necessary to obtain an

OBJECTIVE EXAMINATION OF THE STOMACH

The estimation of the general condition of the patient is of the greatest value. Rapid loss of strength would indicate a gastric carcinoma, whereas a good appearance would contra-indicate it; still extensive emaciation may accompany chronic gastric catarrh and dilatation of the stomach. One should notice the attitude of the patient, the character of his complaints, the expression of his face and of his eyes in order to discover whether he is of nervous temperament.

Inspection is generally of little value. Only in cases of considerable dilatation does the stomach appear as a distended bladder extruding the thin abdominal wall.

Palpation. — On palpation one should be on the watch for

1. *Tenderness*, which accompanies many of these disorders. Only severe localized pain which is increased on pressure would indicate ulcer.

The pain of ulcer of the stomach is especially characterized by the fact that it is produced by the contact of the ulcer with the acid contents of the stomach and increases after food has been taken. When the stomach is free from food in cases of ulcer, the patients are as a rule free from pain. The pain begins soon after eating, increases gradually, and reaches its highest point from 2 to 3 hours after the introduction of food. Vomiting the acid stomach contents produces a diminution of the pain. It is often possible to determine the *location of the ulcer* from the pain, which changes with different positions of the body; patients with an ulcer of the pylorus are relieved of pain by lying on the left side, an act which takes the food away from the pyloric end; should the pain be situated on the posterior wall of the stomach, an abdominal posture will relieve it. *Ulcer of the duodenum* is recognized by the fact that the pain *begins* only *after* the food contents have passed into the small intestine in from 2 to 3 hours after ingestion.

2. *The presence of a tumor.* Only when a tumor can be felt ought we make a diagnosis of carcinoma. Tumors of the stomach move only slightly on respiration, whereas hepatic tumors move up and down distinctly. Should *no* tumor be felt when cancer is supposed to exist, then the diagnosis of carcinoma should be made tentatively; for one may be present on the posterior wall which can not be felt by palpation.

Tumors of the stomach are mostly carcinoma when they occur in old people. Still, there are practically many important exceptions, as: 1. *hypertrophy of the pylorus*, which may feel like a tumor about the size of a pigeon's egg. The diagnosis of this disorder is confirmed by the presence of the signs of ulcer, the long duration of the symptoms, the, as a rule, good preservation of the body, and the failure of the tumor to grow in size. Still, when accompanied by an extensive dilatation, the differential diagnosis may be most difficult. 2. *Perigastritis in chronic gastric ulcer.* Chronic ulcer of the stomach leads frequently to a diffuse infiltration and thickening of the tissues about the ulcer, which may give rise on palpation to the sensation of a tumor. Here, also, a fairly well nourished condition of the body, the long duration of the disease, the preceding symptoms of ulcer, and the structure (consistency) of the tumor may lead to the diagnosis.

Percussion. — By its means the size of the stomach is determined, though simple percussion, as a rule, gives untrustworthy results, because the surrounding intestines give rise to the same percussion note (Fig. 24).

The stomach is so situated in the abdominal cavity that $\frac{2}{3}$ of its volume lie on the left side of the median line, while $\frac{1}{3}$ is on the right of that line. The fundus lies in the concavity of the left vault of the diaphragm, the cardia at the level of the 11th to 12th dorsal vertebra; the small curvature and the pylorus are covered by the liver. The pylorus is situated on the right sternal line at the level of the apex of the xiphoid process. The lower limits of the stomach extend to from 2 to 3 finger breadths above the umbilicus. The space, within which the tympanitic note of the stomach is heard, over the thoracic wall, is called the *semilunar space;* the boundaries of the semilunar space are liver, lung, spleen, and arch of the ribs.

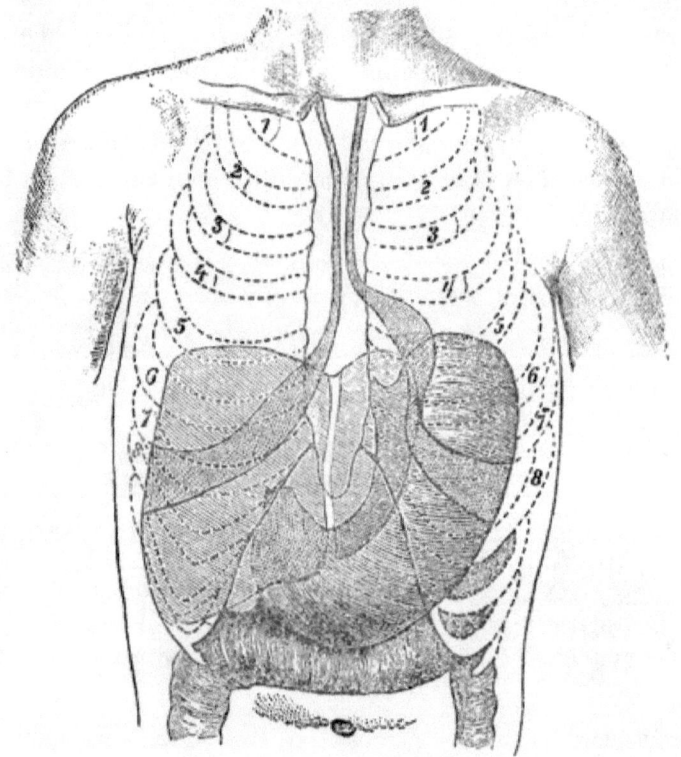

FIG. 24. -- RELATIVE POSITION OF STOMACH, LIVER, AND COLON.

Valuable results are obtained by percussing the stomach when, after its being empty, several glasses of water are taken one after the other during definite intervals. In such a case, on percussing, a gradually increasing area of dulness is obtained corresponding to the drinking of each glass and by which the lower border of the stomach may be distinctly recognized.

The most reliable results in reference to determining the size of the stomach are to be obtained by

Ballooning the Stomach

This ought to be done in all cases in which a suspicion of a dilatation of the stomach is entertained. Should there be a suspicion of the presence of an ulcer, distending the stomach with gas should not be performed, as it would be extremely hazardous.

The operation is performed by administering to the patient when the stomach is empty a teaspoonful of tartaric acid, and following this with the administration of a teaspoonful of sodium bicarbonate dissolved in a little water. Carbonic acid gas is thus generated within the stomach, and as a result the contours of the stomach in many cases become distinctly visible within the abdominal wall. In other cases where the stomach is not thus forced out so visibly, percussion of the organ easily differentiates it from the surrounding viscera. Should the disease be one which would indicate the introduction of a stomach-tube, the stomach may be directly blown up by the careful use of an air-pump.

The stomach is the seat of *dilatation* when its lower border reaches the umbilicus.

It should be observed, however, that there are many people who have naturally an unusually large stomach, which, nevertheless, performs its functions well (*megalograstia*). *Dilatation of the stomach* only refers to those cases of gastric enlargement which are produced by pathological causes and which give rise to disease symptoms. A descent of the lower border of the stomach may be produced by a dislocation of the entire organ; in these cases the transverse colon, as a rule, is also shoved downwards and the other abdominal viscera, such as the kidneys, are displaced, owing to the relaxation of the peritoneal folds (*enteroptosis, Glénard's disease*).

Recently the size of the stomach has been determined by the introduction into the organ of an electric light, which illuminates it so that its contours may be mapped out (*gastrodiaphany*). This method has also been used for diagnosing the presence of a tumor of the stomach. Its use, however, necessitates an expensive instrument and a very strong electric current.

By the use of these methods of examination a differential diagnosis may be made in very many cases.

In several cases the determination of a diagnosis offers many data for the necessary therapeusis; *e.g.*, ulcer and dilatation. In a large class of cases, nevertheless, the treatment depends on the cause of the conditions of the acidity of the stomach. An anatomical diagnosis can only determine this in a few cases, because many cases *of the same form* of disease must be treated differently according as the acidity of the gastric juices varies.

The treatment of many diseases of the stomach, especially of catarrhs and neuroses, is chiefly a dietetic one. But the regulation of the diet is directly dependent upon the conditions of acidity. Patients with a hydrochloric hyperacidity bear meat extremely well, milk also in most cases, but digest the carbohydrates badly and only partially the fats. Patients with an anacidity digest fat and carbohydrates well as a rule if fermentation is checked, whereas meat in large quantities distresses them easily. The medicinal treatment — whether it should be an acid or an alkali — is directly dependent on the knowledge of the acidity of the stomach. Finally, the indications for washing out the stomach may only be first determined by an examination of the stomach contents, as, for instance, in organic hyperacidity.

In all those cases where the anatomical diagnosis presents no sufficient data for the selection of a mode of treatment, we should undertake

THE ANALYSIS OF THE CONTENTS OF THE STOMACH

Methods and normal relation. In the morning, when the stomach is entirely empty, a soft stomach-tube should be introduced, and by means of a small aspirating bottle some of the contents is aspirated. The healthy stomach before breakfast contains nothing or but a little slightly acid fluid. Accordingly the patient should be instructed to drink a pint of milk and to eat some bread (milk test meal). Two hours thereafter the stomach-tube is again introduced, and the contents of the stomach are aspirated. These should be filtered, and the filtered solution examined as to its acidity. The introduction of the tube is easily done without much trouble to the patient; still, a certain skill is required in introducing it which may be acquired by practice. Should the retching which follows the attempt at introduction be excessive, a 10 per cent. solution of cocaine may be applied by a brush to the pharynx about 10 minutes before operating. Introducing a tube in an empty stomach is not necessary in all cases, but is a desirable procedure in cases of hyperacidity.

Instead of the milk test meal, the patient may take: 1. *Ewald's* test meal, composed of a roll and a cup of tea. This should be aspirated from the stomach for test purposes in three-quarters of an hour. 2. The *Leube-Riegel* test meal, which consists of barley-soup, 150 to 200 grammes of beefsteak, 50 grammes of bread and a glass of water, aspirated in 4 to 5 hours. 3. Should an accurate determination of whether *lactic acid* is formed in the stomach be required, it is best to use *Boas's* suggestion of giving, the night before the stomach is to be aspirated, a meal composed of oatmeal (1 tablespoonful of oatmeal, 1 teaspoonful of salt, and 1 quart of water, boiled together). The following morning an aspiration of the stomach should be made.

Instead of aspirating we may gain sufficient of the stomach contents by pressing upon the epigastrium, and having the patient at the same time attempt to vomit (*Ewald's* expression method).

The contents are filtered and the filtered solution is examined:—

1. With litmus paper.

2. For *free acid*. Pour a few drops of a watery solution of *methyl-violet* into a watch glass full of the stomach contents. If but a little hydrochloric acid be present, the solution becomes slightly blue. Another test is to add a few drops of a weak solution of yellow tropäolin; in the presence of free HCl the solution becomes more or less red. A third test is with *Günzburg's* solution, composed of *phloroglucin*, 2 grammes; *vanillin*, 1 gramme; *absolute alcohol*, 30 grammes (to be preserved in a dark bottle), of which one drop is added to several of the gastric fluid in a porcelain dish which is put over a Bunsen burner and slightly warmed; if there be any hydrochloric acid present, red streaks composed of the finest crystals develop.

3. For lactic acid with *Uffelmann's* solution, which is composed of 10 c.c. of a 1 per cent. solution of carbolic acid, 1 to 2 drops of iron perchlorate, which makes a blue-violet color. To this

reagent, which may be placed in a test tube, a few drops of the gastric fluid are added; if a large quantity of *lactic* acid be present, the fluid becomes yellow. In the presence of a large quantity of hydrochloric acid this test will fail, so that it would be necessary to shake 10 c.c. of gastric fluid and 50 c.c. of ether in order to obtain the lactic acid.

4. For *peptone* with the *biuret* reaction; a specimen of gastric fluid is rendered strongly alkaline with sodium hydrate, a 1 per cent. cupric sulphate solution is added drop by drop, and if peptone be present a distinct red color will appear.

5. It is necessary to determine the *total* acidity. For this purpose 10 c.c. of gastric fluid are titrated with deci-normal sodium hydrate solution. It is advisable to use a graduated burette. from which the deci-normal sodium hydrate solution may be dropped, and a pipette with which 10 c.c. of gastric fluid may be put into a glass beaker. The gastric fluid in the beaker is diluted with distilled water until it is nearly colorless, and 2 drops of an alcoholic solution of *phenolphtalein* are added. This solution becomes red in alkaline solutions, while it remains colorless in neutral and acid solutions. After this, one should carefully drop from the burette the sodium hydrate solution into the diluted gastric fluid until the mixture becomes of a faint rose tint. The amount of deci-normal sodium hydrate used indicates the amount of acid the stomach fluid contained; for example, to produce the rose tint in the 10 c.c. of gastric fluid 5.8 c.c. of sodium hydrate were used; therefore in 100 c.c. of gastric fluid 58 c.c. would have been used; the total acidity is hence 58. But since 1 c.c. of deci-normal sodium hydrate solution contains 0.004 g. of NaHO, the 58 acidity means that it would require 58×0.004 g. NaHO to neutralize the amount of acid present. Therefore $58 \times 0.004 = 0.232\% =$ total acidity, or 2.32 per thousand NaHO. Should this result be desired in terms of hydrochloric acid, we can utilize the formula $NaHO : HCl :: 40 : 36.5$, from which the acidity $58 = 0.212\% = 2.12\%$ HCl.

The determination of the total acidity is the most important feature of the examination. If the stomach was washed out thoroughly until clear fluid is returned, before the test meal, then the acid present after the test meal may be considered to be, for the greatest part. hydrochloric acid, even if the methyl-violet reaction is negative. The blue reaction of the methyl-violet is interfered with, as well as are the other color reactions, should there be a large quantity of albuminoids in the fluid examined. The color reaction should be controlled by the biuret test; should this be positively present, then it is proven that hydrochloric acid was effective in changing albumin to peptone. From this result the conclusion is satisfactory that the *absence of the hydrochloric acid color reaction is no proof that there is no hydrochloric acid in the gastric fluid.* On the other hand, the presence of the color test indicates, with sufficient certainty, that hydrochloric acid does exist in the fluid.

Should the stomach not have been washed out before the test meal, then the total acidity represents the combined hydrochloric and other organic acids (lactic, acetic, and butyric). By means of the methyl-violet, the biuret, and the Uffelmann tests, the valuable conclusion whether much organic acid is present is reached. Should it be — in especially important cases — desirable to determine quantitatively the amount of hydrochloric and lactic acids, complicated chemical methods must be employed.

The gastric fluid obtained two hours after a test meal reacts in normal cases. Acid and litmus gives a positive reaction to the hydrochloric acid test, as well as to the biuret test, but no reaction to the lactic acid test.

The total acidity varies between 50 and 65 (c.c. $\frac{1}{10}$ normal sodium hydrate solution to 100 c.c. of filtered solution) = 0.18 to 0.24 % of hydrochloric acid.

The hydrochloric acid reactions are absent in gastritis and in carcinoma of the stomach; they are often markedly increased in ulcer of the stomach and in nervous dyspepsia.

The failure of the hydrochloric acid reaction is by no means calculated to disprove the presence of carcinoma of the stomach; still a marked presence of this reaction would contra-indicate, in most cases, the existence of that disease.

The proof of the presence of a large quantity of lactic

acid would strongly indicate carcinoma. The test, however, is only conclusive providing the stomach has been well washed out, and the test meal is deficient in food containing lactic acid (*Boas's* test meal, see p. 86).

In addition to the acids, the stomach contains the digestive ferments, *pepsin*, which peptonizes the albuminoids, *rennet ferment*, which coagulates milk, and their precursors, pepsinogen and labzymogen. The examination for these ferments has as yet no essential diagnostic significance, as they are present in most cases of stomach disease.

The consideration of the artificially obtained gastric fluid (the question of the relation between the amount thus obtained and the amount of food) justifies the conclusion as to how much food the stomach has passed into the intestine (the *motor activity* of the stomach). From this not infrequently a judgment as to the improvement or non-improvement of a case may be formed. According to *Leube* every stomach is deemed insufficient from which remnants of food may be obtained by aspiration seven hours after a meal.

A better method of determining the motor activity of the stomach is by means of the *salol test*. Salol passes from the stomach into the intestine unchanged, and is split up only by the alkaline secretion of the intestine into salicylic acid and phenol; the reaction of salicylic acid in the urine (a violet color on the addition of ferric chloride) indicates that salol has passed into the intestines. Two (2) grammes of salol should be given in a wafer at a meal. In the normal person, the reaction for salicylic acid will be obtained in the urine in ¾ to 1 hour afterwards, that is, urine thus obtained will show a violet color to ferric chloride. In motor weakness of the stomach this reaction will appear from 2 to 5 hours afterwards. In addition a distinct reaction may be obtained in cases of good motor function at the utmost up to 24 hours afterwards, while in motor insufficiency it will persist for 2 days.

It is desirable to determine more definitely in a few individually important cases, for instance in stricture of the pylorus, the motor activity of the stomach. For this purpose 100 c.c. of pure olive oil are poured into the stomach which has been previously washed out, and after 2 hours an aspiration of the gastric fluid is made (*oil test*). The normal stomach will pass 50 to 75 c.c. of the 100 c.c. of oil that had been given into the intestine; in very many diseased conditions the amount of oil obtained from the stomach is more or less diminished.

Chief Symptoms of the Most Important Diseases of the Stomach

Acute gastritis. — Severe dyspeptic symptoms are prominent. It is caused chiefly by errors in diet. Vomiting is an accompaniment, together with diminution or absence of hydrochloric acid. The region of the stomach is tender. Not infrequently there is headache and some fever. The termination is mostly favorable, still a transition into a chronic gastric catarrh is possible.

Ulcer of the stomach (*ulcus ventriculi*). — Is manifested by *localized pain*, with vomiting after the ingestion of food, the vomited material being often strongly acid. Vomiting of blood also occurs (*hæmatemesis*). The condition of nutrition of the body is somewhat below par. The acidity is most always increased, still when anæmia is present it may be diminished.

Carcinoma of the stomach. — There is a palpable *tumor* of the organ. Vomiting of a contents of a coffee-ground appearance occurs; the patient is distinctly cachectic. The reaction to the test for hydrochloric acid is almost always absent, but lactic acid in the stomach contents is increased. The duration of the disease is between 2 and 3 years.

The dyspeptic symptoms which appear in old persons unaccompanied by a palpable tumor and by vomiting, and which in spite of a rational system of treatment persist and lead to excessive emaciation, may be mistaken for carcinoma of long standing.

Dilatation of the stomach. — At more or less long intervals vomiting of large quantities takes place. The lower border of the distended stomach lies below the umbilicus. The bowels are constipated, the amount of urine is small, and the skin is dry. Emaciation is noticeable.

The diagnosis should at the same time embrace the causative factor of the disease: whether the disorder is due to stricture of the pylorus or to atony of the muscular structure of the stomach. The cause of stricture is to be found in, either a cicatricial formation (due to an ulcer or to the inflammation resulting from caustic substances taken into the stomach) or in a carcinoma. The stricture is sometimes produced in rare instances by the compression of a tumor, by twist or bend of the duodenum caused by peritoneal adhesions, by a floating kidney, etc. The atonic dilatation is the result of direct distention, due to excessive drinking and gluttony, and of chronic gastric catarrh.

Chronic gastritis. — Palpation and percussion of the stomach demonstrates, outside of the abnormal sensitiveness of the organ, no foundation for the severe dyspeptic symptoms. If proof of taking injurious substances into the organ can be deduced, the diagnosis of the disease is probable. The stomach contents contain much mucus, and the acidity is diminished (still in a few cases it is increased).

Nervous dyspepsia. — The symptoms which may be objectively demonstrated give no indication of the severity of the disorder. There is often diffuse pain over the stomach, mostly independent of the ingestion of food and the position of the body, and often hydrochloric hyperacidity. When a neurotic condition can be demonstrated and neurasthenic symptoms are present, the diagnosis is easy; it may also be made if the causes which lead to gastritis can be excluded.

Diagnosis of the Diseases of the Intestines and the Peritoneum

The diseases of the intestines and of the peritoneum are recognized outside of the consideration of the general condition of the patient by an examination of the *fæces* and of the *abdomen*.

Fæces

The normal person has one or two movements of the bowels in twenty-four hours. The stool is formed and of the consistency of thick mush. *Thin* or *fluid* stools constitute *diarrhœa*. (See p. 93.) The color and amount of the fæces depend upon the food. When meat has been the chief article of diet, the amount is small, the color is brownish, and the fæcal mass is firm; when chiefly bread and potatoes are eaten, it is more voluminous, soft, and yellowish brown; when milk is exclusively taken, it is yellowish white, and tolerably firm.

The ordinary color of the fæces is caused partly by the reduced coloring matter of the bile (hydrobilirubin). The reduction is caused by the bacteria in the intestines.

Abnormally colored stools: black stools are the result of hæmorrhage (see below), or are occasioned by medicines, such as iron, bismuth, which form respectively sulphide of iron and sulphide of bismuth. *Green stools* are produced by mercury, especially by *calomel*, which forms mercuric sulphide at the same time as it does biliverdin, or by the passage through the intestine of unchanged bile, which occurs in the *summer diarrhœa* of children. *Grayish-white stools* occur in fæces free from bile, and when the fæces contain much fat.

Fatty stools are grayish-white, mucilaginous, and smell badly; they occur as the result of the occlusion of bile from the intestine (icterus), and in addition chiefly from chronic peritonitis and intense anæmia.

Bloody stools appear as blackish-brown fæcal masses looking like *tar*, and are present in *enteritis*, tumors of the intestine, embolism of the mesenteric arteries, in typhoid fever, and in purpura. Hæmorrhages from piles give rise to a bright-red color of the stool.

Purulent stools. Should the stool contain only pus, the perforation of a peritoneal exudate into the intestine may be diagnosed; when only mixed in small quantities in the fæcal mass, and associated with fluid and at times bloody stools, it would indicate an *ulcer* of the large intestine. Bloody purulent stools, frequently in small amounts and accompanied by severe tenesmus, indicate *dysentery*.

Mucous stools. Pure mucus denotes *catarrh* of the *rectum*, as does the encapsulation of the scybalous stool with mucus.

Small particles of mucus intimately mixed like grains (often like sago-grains, or only microscopically demonstrable) in the formed stool, indicate the presence of a catarrh of the small intestine. Sometimes these particles are colored with bile. Mucous formation in the shape of tubes (mucous casts) occurs in *mucous colic* (a secretory neurosis of the large intestine). *Particles* of the *tissues* in the stool indicate an ulcerative process.

Microscopical examination may reveal muscular fibres, fat (in masses, in particles, or as crystals), crystals of triple-phosphates, *Charcot-Leyden* crystals (often with entozoa). Leucocytes are only of diagnostic importance when very numerous. Mucous particles may also be seen. In chronic catarrh one will often find desquamated epithelium (homogeneous, non-nucleated, and spindle forms).

Constipation occurs in people of sedentary habits, and is often the result of deficient physical exercise, of excessive eating of meat, or of deficient ingestion of fluids; it occurs in patients who are compelled to lie in bed, as the result of the unaccustomed rest; frequently it is present in cases of pregnancy and in cases of hysteria; also in cases of dilatation of the stomach, in appendicitis, in cases of obstructive venous circulation, such as results from uncompensated heart lesions, etc. Still diarrhœa also occurs in these cases. When alternating with attacks of diarrhœa, it indicates chronic intestinal catarrh. Long-continued constipation occurring at the same time with an *inability* to *pass gas* should awaken the suspicion of an intestinal obstruction (ileus, see p. 95).

DIARRHŒA

Thin copious stools are produced

1. By increased peristalsis, which is caused by (*a*) nervous shocks, such as fright, fear; (*b*) cold; (*c*) irritating substances which appear in the intestinal tract, especially those that ferment and decompose (gastro-intestinal catarrh, summer diarrhœas); (*d*) specific bacterial influences, probably likewise producing irritating chemical changes (colitis, dysentery, and cholera); (*e*) ulcerative processes in the mucous membrane (exfoliative ulceration as in chronic enteritis; typhoid, tubercular, and syphilitic ulcerations); (*f*) the circulation of toxic compounds in the blood, as in uræmia and probably in carcinoma.

2. By the fact that owing to the degeneration of the epithelium of the intestine, an adequate resorption of chyme does not take

place. The series of causes just mentioned may also lead to this deficient absorption. To this category belong especially (a) amyloid degeneration of the intestinal mucous membrane; (b) the obstruction in the portal circulation (cirrhosis of the liver) and in the general venous circulation (non-compensated cardiac disease). The obstructive conditions lead again to catarrh, and belong therefore in part to the first group.

3. By the transudation of watery fluids into the intestine; all infectious and inflammatory states fall under this group.

Diarrhœa of *short* duration, mostly without significance yet leading quickly under certain conditions to severe symptoms, occurs in cases of nervous or psychic excitement, in exposure, in acute gastro-intestinal catarrh, and is also infectious, especially when occurring in children. It occurs as a symptom of cholerine, summer diarrhœa, and cholera nostras.

Diarrhœa of *longer* duration, chronic diarrhœa, occurs in portal obstruction; in interference with the general return circulation as in some forms of heart disease; in the various intoxications, *e.g.* uræmia, as well as in certain sub-acute infectious diseases, such as typhoid fever; in general this kind of diarrhœa justifies the diagnosis of a grave disorder of the intestine (either a chronic enteritis, a dysentery, a tubercular or syphilitic ulceration, or an amyloid degeneration). Tuberculosis of the intestine is only to be diagnosed with any degree of certainty when it can be proven to exist in other organs of the body, and amyloid degeneration only when one of its etiological factors, such as phthisis, syphilis, malaria, abscess, etc., is present. The situation of the ulcerative process may often be determined by the character of the mucus or pus mixed with the fæcal discharges (see above). The following diseases have characteristic stools: *typhoid* fever, where the stool has a "pea-soup" appearance; *dysentery*, where it is mucous and bloody; and *cholera*, where it looks like "rice-water."

Abdomen

Normal relations. The abdomen of the healthy person is moderately arched, moves up and down during respiration, is usually soft, easily depressed, and nowhere painful on palpation; offers no resistance anywhere to pressure and is tympanitic on percussion excepting when the colon contains large fæcal masses, when the percussion note may be dull.

Retraction of the abdomen. The abdomen is *scaphoid* in shape and retracted in spasm of the intestine such as occurs in colic and in meningitis, or when it is entirely empty as in inanition, in dilatation of the stomach, and in stricture of the œsophagus.

Distention of the Abdomen

The abdomen may be highly arched, and sometimes tense as a drum. These conditions are caused by

I. An accumulation of gas in the intestines (**meteorism**, tympanites); determined by the general tympanitic resonance on percussion, the absence of fluctuation. Moderate tympanites occurs in typhoid fever, in chronic intestinal catarrh, and in obstruction in the portal circulation.

A *high degree* of tympanites occurs with (1) intestinal obstruction (*ileus*). The diagnosis of ileus is rendered certain when, in addition to extensive tympanites, there is the facies of collapse, small frequent pulse, and *fæcal vomiting*. After the diagnosis of ileus is made, the following data must be determined : 1. The location of the obstruction; is it situated in the small or large intestine? 2. The nature of the obstruction, which may be an invagination, a twist (volvulus), a strangulated hernia, a constriction due to a peritoneal band, a peritoneal abscess, a cicatrix of an old intestinal ulceration, a malignant tumor, a foreign body such as a gall-stone.

The hernial canals and the rectum should always be examined without fail. Obstruction of the small intestine is accompanied by active visible peristaltic motion in the intestines, by a large amount of indican in the urine (Chap. VIII.), and by violent general symptoms; in obstruction in the large intestine, a great amount of distention of the portion of the colon above the constriction is observed, the urine contains but little indican, and the general symptoms develop more slowly. The diagnosis of the nature of the obstruction is made in many cases from the history of the case and by manual examination, but can not often be absolutely settled.

A high degree of tympanites occurs with (2) acute general peritonitis, where every movement of the patient is attended by pain, as is every touch on the abdomen; in addition there are present, vomiting of bile, a small frequent pulse, and the *facies hippocratica*.

Even in localized peritonitis an irritation of the entire peritoneum may result, so that the symptoms of a general peritonitis may be closely imitated. The diagnosis is determined by the proof of the presence of an exudate.

II. By a collection of free fluid in the peritoneal cavity (ascites). In these cases the abdomen, when the patient is lying down, is distended at its sides and flattened in the middle; above there is tympanitic percussion, while dulness is elicited in its dependent parts. Dulness on percussion is bounded above by a horizontal line and is changed by a change in the position of the patient. On tapping the side of the abdomen with the fingers, *fluctuation* is often elicited.

If the patient is placed on his side, a tympanitic percussion note is obtained over the side which is uppermost; should the position be changed to the recumbent one, dulness on percussion will be found on the side which had just shown a tympanitic note.

Should ascites have been diagnosed, the following possibilities should be regarded: —

1. Ascites, together with general œdema, occurs in cardiac and renal diseases, but the ascites is only a secondary development to the original œdema, and is of itself not of essential importance.

Still, in consequence of a long-continued ascites, œdema of the legs may follow as a result of the pressure of the fluid on the veins and of the anæmia.

2. Ascites may be unaccompanied by general œdema, or only by a secondary œdema of the legs. In this case there is present either: —

(*a*) *Obstruction to the portal circulation* by reason of hepatic disease or occlusion of the portal vein; under these conditions the ascitic fluid contains very little albumin, and varies in *specific gravity* from **1006** to **1015**.

(*b*) *Chronic peritonitis.* In this case the ascitic fluid contains a larger percentage of albumin, and its specific gravity is 1018.

The specific gravity is measured by a urinometer, the fluid to be examined being at the temperature of the room. When taken at the temperature of the body, it is lower than under the former conditions; it is 1° of the urinometer lower for every 3° C. elevation above the temperature of the room. From the specific gravity the amount of albumin may be proximately determined, in this manner, according to *Reuss's* formula, $A = \frac{3}{8}(S - 1000) - 2.8$, where A represents the amount of albumin in per cent., S, the specific gravity.

To return to (*a*), which we might call congestive ascites, it occurs chiefly in diseases of the liver, and especially in *hepatic cirrhosis*. Accompanying the ascites is a distention of the veins of the abdomen, especially those about the umbilicus (forming the *caput medusæ*); in addition the spleen is enlarged, and the stomach and intestines are the seat of a catarrhal inflammation. From the history will be elicited that the patient has indulged excessively in alcohol. In rare instances syphilis produces a peculiar variety of hepatic cirrhosis (*hepar lobatum*), which proceeds without ascites, and always runs a chronic course.

The other diseases of the liver, especially carcinoma and syphilis, give rise only seldom to ascites, and are diagnosed by the absence of the characteristic signs of cirrhosis, by palpation, and by the history.

Occlusion of the portal vein occurs very rarely, and is then produced by tumors of the stomach, of the pancreas, etc., or by a thrombosis of the vein.

(*b*). *Chronic peritonitis* is produced by carcinomatosis or by tuberculosis of the peritoneum, is accompanied by general cachexia, and the diagnosis is made positive when either carcinoma or tuberculosis is discovered in one or more of the other organs.

In chronic peritonitis the ascites may be encapsulated by peritoneal adhesions. In such a case the change of the position of the patient may not give rise to change in the percussion note. *Fric-*

tion sounds may be heard and friction be felt. Occasionally an infiltration of the abdominal wall around the umbilicus is found (*periomphalitis*).

III. *By tumors,* in which case the abdominal distention is often asymmetrical, being most prominent, not infrequently, at the site of the organ affected. Over the distended area dulness on percussion is present. Tympanites may occur at the same time. A diagnosis of tumor can only be certain when a *tumor can be felt*.

We should be on our guard not to confound tumor with an accumulation of fæces in the intestine. In the latter case the mass is doughy in consistency, easily indented, and disappears after a free catharsis.

Tumors may arise in the liver, the spleen, the kidneys, the intestines, the stomach, the omentum, and only seldom in the vertebra and the pelvic bones, in the aorta (pulsating aneurysm), and in the female genital organs.

A tumor in the *ileocæcal region* of the size of an egg, an orange, or even up to a saucer, if tender and combined with fever, vomiting, and tympanites, denotes the presence of an *appendicitis* or perityphlitic exudation.

Tumors of the ovary and the pregnant uterus will show dulness on percussion over the lower half of the abdomen; the upper boundary of dulness is convex upwards, and in the lateral parts of the abdomen in the recumbent position it is tympanitic; there is no change in percussion note.

To determine from which organ the tumor arises is often extremely difficult. To assist in the diagnosis, various aids are called into requisition, such as ballooning the stomach or filling it with water, or by doing the same to the colon by means of a rectal tube.

IV. By the *escape of gas* into the peritoneal cavity. Gas in the abdominal cavity will always rise to the top, so that when the patient lies on the left side, the area of liver dulness will disappear, and when on the right side, no splenic dulness can be elicited. Gas within the peritoneal cavity is the pathognomonic sign of a *perforation peritonitis*, which is almost invariably fatal.

Perforation of the stomach may be produced by a long-standing ulcer of the stomach, especially after an intense bodily strain, or after a full meal; perforation of the intestine may result from an ulcerative process in that viscus, especially occurring in the remission stage of typhoid fever, and after there has been a hæmorrhage from the bowel. Another frequent cause of perforation peritonitis is the perforation of the vermiform appendix by a fæcal concretion.

The *disappearance of liver dulness* has a fatal significance only when *at the same time a change in position of the patient produces a change of percussion note;* changing the patient's position, however, occasions intense pain, and should be done only in cases of urgent necessity. Perforation peritonitis is also made apparent by the signs of great collapse, as well as by the local signs.

The liver dulness is absent frequently in ordinary tympanites of moderate amount, such as occurs in constipation; in addition, it may be absent or much diminished in diseases of the liver, which lead to atrophy, such as acute yellow atrophy and cirrhosis. In acute atrophy of the liver a progressive diminution in liver dulness may be seen taking place from day to day. It is diminished in cases of extensive pulmonary emphysema, where the distended lungs overlap the liver area.

Chief Symptoms of the Most Important Diseases of the Intestines and Abdomen

Acute intestinal catarrh: suddenly developing diarrhœa, with colic and tenesmus. In severe cases the final evacuations contain much watery mucus. It is often preceded by vomiting, and is sometimes accompanied by fever. There is almost always more or less depression. *Chronic intestinal catarrh* is characterized by a diarrhœa, which lasts for months, consisting of mucus, pus and bloody stools, alternating with constipation, slow emaciation. Specific causes, such as tuberculosis, syphilis, dysentery, and carcinoma, which may likewise be the cause of a chronic catarrh, should be excluded in making a diagnosis.

Mucous colic (formerly known as *enteritis membranacea*) is a rare form of intestinal affection, occurring in neuropathic individuals, and characterized by a diarrhœa accompanied by attacks of colic, the stools containing tape-like, cylindrical, pseudomembranous casts of coagulated mucus.

Carcinoma of the intestine: cachexia, and a palpable tumor in the abdomen, which is often very movable. The intestine may be moved with the tumor. It is frequently associated with signs of intestinal obstruction (see p. 95) and with hæmorrhages from the bowels. Tumors of the rectum can usually be felt from the anus.

Ileus. (See p. 95.)

Acute general peritonitis. There is extensive tympanites, vomiting of bile (spinach-green vomiting), at times singultus. The abdomen is intensely tender, and when due to a perforation the liver dulness gives place to tympanitic percussion, owing to the contained gas rising to that region. The facies of collapse are well marked, the pulse is frequent and small, and respiration is shallow and rapid.

Localized peritonitis. The general symptoms approximate more or less those of general peritonitis, though as a rule they are much less intense. By palpation a localized exudation can be felt. The rectum should be examined by the finger for this purpose.

Perityphlitis, appendicitis. A painful infiltration or exudation in the ileocæcal region may be felt. On percussion it is dull. There is vomiting, tympanites, constipation, and fever. The pulse is good, as a rule. It ends frequently in spontaneous absorption, though rupture of the abscess occurs; operation is not infrequently necessary.

Chronic peritonitis. (See p. 97.)

DIAGNOSIS OF DISEASES OF THE LIVER

The important data to be obtained in simple jaundice (icterus) are: mistakes in diet, previous gastric catarrh, fright, anger; when severer symptoms are present the following: a previous attack of jaundice, especially of biliary colic (gall-stone); a history of alcoholism (cirrhosis); of companionship with dogs (echinococcus); of syphilis. One should determine also whether the etiological factors of amyloid degeneration are present and whether there has been any history of poisoning, for instance, with phosphorus.

Diseases of the liver are recognized in many cases by the presence of *icterus*, which is first observed in the conjunctiva and finally over the skin of the entire body. The urine is of the color of dark beer, shows *Gmelin's* reaction (see Chap. VIII.); the fæces are *grayish-white*, like *clay*. The following types are recognized:—

1. *Icterus simplex* (catarrhal jaundice), which is associated with mild symptoms, headache, fatigue, malaise, itching of the skin, and a slow pulse. It arises as the result of the occlusion of the ductus choledochus by a catarrhal inflammation of its mucous membrane. It lasts but a few weeks. The prognosis under proper treatment is favorable.

2. *Icterus gravis* occurs with severe symptoms of disease, with emaciation, frequently with fever, mental confusion, delirium, and severe pains in the region of the liver. It may be produced by passage of a gall-stone or impacted gall-stone, by liver abscess, by echinococcus of the liver, by carcinoma of that organ, and by acute yellow atrophy.

Concerning icterus with polycholia and without polycholia, see p. 8. The clinical consideration of cases of severe jaundice may present many prominent symptom-combinations which serve to make the diagnosis easy; thus icterus *combined with cachexia* indicates *hepatic carcinoma;* when combined *with ascites* it indicates *cirrhosis; icterus* occurring with *attacks of colic* indicates *biliary calculi;* when associated with *chilly sensations, abscess of the liver* is probably present. Of course these symptom-combinations have only a limited diagnostic value; they should always be corroborated by the other signs of the individual diseases elicited only by a careful examination. Thus, the picture of a jaundice due to gall-stone may be simulated by a carcinoma of the duodenum and that of cirrhosis of the liver by a chronic peritonitis, etc.

There are some diseases of the liver in which jaundice appears only late in the course of the disease or *not at all*. They are amyloid liver, fatty liver, congestive liver (due to impeded venous circulation), carcinoma of the liver, syphilitic liver, atrophic cirrhosis, and echinococcus. The attention of the clinician is drawn to the liver chiefly by the patient's symptoms of pressure and pain in the hepatic region, sometimes by the ascites, and often only after a casual physical examination of the organ.

A differential diagnosis is made from the history of the case and the general condition of the patient, and by the physical signs obtained by *percussion* and *palpation* of the organ.

The liver (compare Fig. 24, p. 83) lies in the right hypochondrium and its *upper border* in the normal state reaches the inferior border of the 7th rib in the axillary line; in the mamillary line it reaches the inferior border of the 6th or the superior border of the 7th rib; its upper border lies behind the 6th rib at the right sternal line. Its *inferior border* in the axillary line lies between the 10th and 11th ribs, *extends across the costal arch in the mamillary line*, and lies midway in the linea alba between the xiphoid process and the umbilicus; it then curves upwards and reaches the diaphragm at a spot between the parasternal and mamillary lines. On deep inspiration the liver descends a trifle.

Percussion of the liver will show relative dulness in the mamillary line from the 4th rib downwards, which changes to absolute dulness (flatness) at the lower border of the 6th rib. The latter ends in the mamillary line at the free border of the ribs, when the percussion note becomes tympanitic. *Palpation* gives no results in the normal condition of the organ in the mamillary line. But in *hypertrophy* of the liver the edge of the liver may be felt below the free border of the ribs and percussion will show dulness also below the free border.

Liver dulness is *increased* always in hypertrophic cirrhosis, in amyloid degeneration of the liver, in chronic congestion of the liver, often fatty degeneration of the liver, in hepatic echinococcus, carcinoma, and abscess.

Liver dulness may extend beyond the free border of the ribs even when the liver is not increased in size, providing the *diaphragm* is *shoved downwards,* as occurs in pulmonary emphysema, in pneumothorax, and in pleuritic effusions on the right side.

Liver dulness is *diminished* in area in acute yellow atrophy (here the diminution goes on from day to day, and without an increase of tympanites), in atrophic cirrhosis (diminution proceeds extremely slow), *often in tympanites* when the transverse colon lies between the liver and the abdominal parietes; should extensive tympanites coexist with grave general symptoms, and in addition should it be possible to elicit by a change of the position of the patient to the right side a liver dulness which was absent when the patient was on the left side, the diagnosis of air or gas in the peritoneal cavity should be made (perforation peritonitis, see pp. 98. 99).

The border of the liver is *felt* to be *smooth* in chronic congestion, in amyloid degeneration, and in hypertrophic cirrhosis. The border and the upper surface are *rough* (uneven and nodular) in atrophic cirrhosis, in syphilitic hepatic disease, in carcinoma, and sometimes in hepatic abscess.

A special form of liver tumor is the *corset or tight-lace liver* (constricted liver), where a portion of the right lobe lies below the free border of the ribs as an isolated tumor extending from 4 to 6 cm. into the abdominal cavity. Abnormal mobility of the liver is known as *floating liver* (it occurs in women with pendulous abdomens).

Chief Symptoms of the Most Important Diseases of the Liver

Catarrh of the bile-ducts (catarrhal jaundice) is characterized by icterus, accompanied by mild symptoms (see pp. 8 and 101); the liver is often enlarged, is but little tender, and the gall-bladder is often palpable. A favorable termination occurs in from 3 to 5 weeks.

In a few cases it progresses to *chronic* jaundice by reason of the agglutination of the mucous walls of the ductus choledochus and the occlusion of its canal (*cholangitis chronica fibrosa*). As a result gradual slow emaciation occurs and after a few years, death, preceded by coma and convulsions, ensues.

Abscess of the liver gives the symptoms of jaundice, combined with erratic chills, emaciation, severe pains in the region of the liver and in the right shoulder. In solitary abscess a protrusion of the liver surface upwards or downwards is sometimes observed. In multiple abscesses the liver is not infrequently enlarged throughout.

Biliary colic (gall-stone colic) is accompanied by very severe attacks of pain in the hepatic region of varying duration and occasionally by jaundice; vomiting and fever are not infrequent. The diagnosis is rendered positive by the finding of gall-stones in the fæces.

Analysis of gall-stones. Gall-stones are composed of concretions either of *bilirubin* or of *cholesterin*. Cholesterin is tested for as follows: a portion of the calculus is dissolved in hot alcohol and the solution is then filtered; from the filtered solution, when cool, cholesterin forms in rhombic crystals. To continue the test, the crystals are dissolved in chloroform and concentrated sulphuric acid is added to the solution, when a beautiful deep-red color appears, which gradually changes to blue and finally green. Bilirubin is isolated by extracting from the remaining filtrate with warm chloroform, the filtrate having previously been rendered slightly acid with hydrochloric acid, when it is observed in the chloroform solution on the addition of fuming nitric acid (*Gmelin's* test).

Carcinoma of the liver give rise to cachexia, combined with icterus; palpable, nodular tumor in the liver region; usually decided enlargement of the liver without any enlargement of the spleen.

Acute yellow atrophy of the liver is diagnosed when a suddenly developing jaundice occurs associated with severe cerebral symptoms such as confusion, delirium, and coma; with a rapid diminution of liver dulness. Tyrosin and leucin are found in the urine, the excretion of urea is very much diminished, the excretion of ammonia very much increased. It pursues a rapidly lethal course.

Acute yellow atrophy may appear after an initial stage of catarrhal jaundice lasting from 8 to 14 days.

Hypertrophic cirrhosis of the liver is characterized by the presence of jaundice, combined with a decidedly uniform enlargement of that viscus; alcohol and syphilis are the etiological factors. It is associated with a visible enlargement of the veins of the abdominal walls, with a hypertrophy of the spleen, and with gastro-intestinal catarrh; ordinarily it is unaccompanied by ascites.

Atrophic cirrhosis of the liver is diagnosed by the presence of ascites, the ascitic fluid having a low specific gravity; by the distention of the abdominal veins and an appreciable splenic enlargement. There is a slowly developing cachexia and sometimes jaundice. Symptoms of gastro-intestinal catarrh. The history will show excessive drinking of alcoholic beverages, more seldom syphilis or malaria or chronic peritonitis or cardiac disease.

Echinococcus of the liver is only to be diagnosed when by the growth of the cyst the liver becomes enlarged. In well-developed cases fluctuation and hydatid fremitus over the tense elastic tumor may be elicited. Aspiration of the tumor mass yields a fluid in which portions of the echinococcus membranous wall and the hooklets of the parasite may be found on microscopical examination. The fluid does not become turbid, or at best but very little, on heating (see Chap. XII.)

Amyloid degeneration of the liver.—There is a uniform firm enlargement of the organ and a cachexia. Proof of the etiological factors, such as phthisis, syphilis, suppuration, etc., will assist in the diagnosis. The spleen is enlarged and albuminuria and diarrhœa are present.

Chronic congestion of the liver also presents a uniformly enlarged liver. Dyspnœa and cyanosis are features and point to the primary cause of the disease; namely, to cardiac or pulmonary affections.

The Spleen

Enlargement of the spleen is an extremely important sign, upon which depends the diagnosis of many diseases.

The spleen is situated in the left hypochondrium. Under normal conditions *splenic dulness* extends from the 9th to the 11th rib, and from the linea costo-articularis (a line drawn from the left sterno-clavicular articulation to the apex of the 11th rib) to the vertebral column. Should the spleen enlarge, its area of dulness will be increased until it finally extends beyond the left costal arch. When the organ is decidedly enlarged, its edge *may be felt* as a sharp border, especially on deep inspiration. Palpation of the viscus is often painful.

A positive proof of enlargement of the spleen is only possible when the organ can be *felt* by *palpation;* it is often possible to demonstrate an enlargement by percussion, though the results of percussion are frequently deceptive on account of the variable amount of fæces in the colon.

The determination of **splenic enlargement** is indispensable in making a diagnosis of *typhoid fever*, of *malarial intermittent*, and of *splenic leucæmia;* it is desirable to demonstrate it to reach a diagnosis of *amyloid degeneration*, of *cirrhosis of the liver*, and of *hæmorrhagic infarction of the spleen.*

It may be enlarged in all infectious diseases; in addition to typhoid, in pyæmia, pneumonia, etc. Should it be so found with pneumonia, it will remain demonstrable until resolution is completed.

The spleen may be displaced downwards in left-sided pleural effusions, in pneumothorax, and in pulmonary emphysema. It is displaced upwards in tympanites, in ascites, and by tumors of the abdominal cavity. Percussion dulness disappears in perforation peritonitis when the patient lies on the right side, and in general is absent over the splenic region in cases of *floating spleen.*

CHAPTER V

DIAGNOSIS OF THE DISEASES OF THE UPPER AIR-PASSAGES (NOSE, THROAT, LARYNX)

The diseases of the nose, to which attention is directed by the discharge, pain, and, above all, the *occlusion* of the nose, may be recognized by inspection with the nasal speculum and palpation with the nasal probe. The symptomatology of all the diseases of the nose would overstep the bounds of this work, and only those symptoms will be considered which are of importance in internal medicine.

Headache, especially frontal, is a symptom of many diseases of the nose. It may be present in a mild degree in an ordinary cold, usually to be referred, however, to disease of the frontal cavity. Purulent processes in this cavity (empyema of the frontal sinus) may give rise to meningitis; next to inflammations of the middle ear, the nose is probably the most frequent source of meningeal infection.

Neuralgia of the branches of the trigeminus is not rarely seen in disease of the nasal accessory chambers, and has been observed especially in connection with disease of the antrum of Highmore.

Epistaxis demands an examination of the nose; but the constitutional causes must not be lost sight of (chlorosis, leucæmia, anæmia, cirrhosis of the liver, contracted kidney, etc.).

A foul smell in the nose is the sign of *ozæna*. The nasal cavity is broadened (atrophy of the turbinated bones), the mucous membrane thin, pale, covered with numerous crusts.

The peculiar bad odor clings to the dried secretion. The secretion may be expelled through the nasopharynx; therefore ozæna is always to be looked for when there is a putrid expectoration (bronchitis, gangrene of the lung).

Mouth-breathing, which appears in every case of occlusion of the nose and nasopharynx (by hypertrophic catarrh, polyps, deviations of the septum, post-nasal tumors, etc.), is often accompanied by disturbing sequelæ: snoring, restless sleep (nightmare), inflammatory processes of the upper air-passages, disturbances of digestion; in children, after a considerable duration, typical facies (open mouth, dull expression), deformities of the teeth, the jaws, even of the thorax.

Adenoid vegetations, hypertrophy of the lymphoid elements in the nasopharynx (faucial or 3d tonsils), exceedingly common in children, give the most exquisite examples of the disturbances of mouth-breathing. They are a frequent source of deafness and may retard the complete bodily and mental development of a child (*aprosexia nasalis*). May be diagnosticated by the child's facial expression; positive diagnosis by the palpation of nasopharynx with the index finger.

The symptoms usually disappear rapidly after the removal of the vegetations.

Nasal reflex neuroses. — These are reflexes called forth by diseases of the nose. Attacks of migraine and spasms of coughing, epileptic seizures, etc., occasionally disappear after the cure of nasal affections. The best known of these is *nasal asthma:* attacks of bronchial asthma dependent upon an irritation from a diseased nose. An existing asthma is only to be referred to the nose when irritation of a particular spot induces an attack while its anæsthesia (by cocaine) stops the attack. Surgical measures for the relief of the asthma may then be undertaken. The same safeguard of diagnosis should be applied to the other neuroses of the diseased nose.

Throat and tonsils. — The inspection of the throat with depressed tongue shows whether infectious processes are localized there (angina or diphtheria, Chap. II.). At the same time, the presence of a *chronic pharyngitis* should be noted (mucous membrane puffy, reddened, covered with secretion or atrophic, dry, occasionally covered with fine granulations); it is often caused by the same elements that help to produce a chronic gastritis (alcoholism, too much smoking, the eating of very hot or poorly masticated food); it is seen in those working in dust of any kind, in speakers and singers.

Insensibility of the fauces is frequently a sign of hysteria or of advanced alcoholism, and may assist in the diagnosis of a contemporaneous gastric affection as neurotic or alcoholic. *Hyperæsthesia*, excessive gagging movements on touching, in chronic pharyngitis, also frequently in drinkers.

Retropharyngeal abscess is recognized by a fluctuating, bulging mass in the posterior pharyngeal wall; there are severe general disturbance and high fever. Deglutition and respiration may be interfered with. The threatening symptoms disappear upon the emptying of the abscess.

The source of cryptogenetic sepsis should be sought in retropharyngeal abscess and purulent processes in the accessory sinuses of the nose.

General Symptomatology of the Diseases of the Larynx

The symptoms of laryngeal diseases are : —

1. *Changes in the* **voice**. — In the great majority of the diseases of the larynx, the voice is changed. It becomes hoarse, rough, rasping, indistinct (dysphonia). Aphonia (absence of voice, speaking in whispers) is a sign of severe disease of the larynx (ulceration of the vocal cords), or of imperfect closure of the glottis (paralysis of the vocal cords). *Falsetto voice* is usually a purely functional disturbance, a defective formation of the voice, which may be overcome by methodical exercise in speaking.

Ventricular voice. A peculiar rough and rattling voice is produced by the employment of the false vocal cords (ventricular bands) for the true ones, frequently in consequence of the disturbance of the latter.

The *nasal voice* may be *open* or *closed*. The former is due to the absence of the closure of the nasopharyngeal opening in consequence of the paralysis of the soft palate (principally after *diphtheria*) or in consequence of ulcerative processes of the soft palate (generally by syphilis). Test-words for this variety are, for example, pump, mumps. When this type of nasal twang exists, fluids taken into the mouth will return in part through the nose. A *closed* nasal voice arises from total occlusion of the nose (chronic catarrh, polyps, etc.).

Diphthongia (diplophonia) is the simultaneous production of two notes of different pitch in speaking; it sometimes occurs in unilateral paralysis of the vocal cords, as well as in the presence of small tumors on the edge of a vocal cord which, in speaking, are pressed between the vocal cords.

A *threefold splitting of the voice* is a rare phenomenon, produced by pediculated polypi which lie beneath the glottis. In speaking they are first pressed between the vocal cords in the expiratory act and then above the glottis. A prolonged vowel is first pronounced clearly, then hoarsely and diphthongic, and finally clearly again.

2. *Disturbances of* **breathing**. — The laryngeal diseases may lead to dyspnœa through narrowing of the glottis. Laryngeal dyspnœa is *inspiratory*, accompanied by *stridor* (a creaking sound the result of the labored, long-drawn breath). The number of respirations is diminished; all the accessory muscles of inspiration are employed (see p. 125); the sides of the neck, intercostal spaces, epigastrium, are drawn in. Expiration is short, inaudible. Laryngeal stenosis occurs in children most commonly in consequence of diphtheria (croup), but occasionally through simple acute laryngitis (pseudo-croup), because of the narrowness of the child's glottis; in adults it is always a sign of severe laryngeal affec-

tion (acute œdema of the glottis, diphtheria, bilateral posticus paralysis, etc.).

Inspiratory dyspnœa and stridor are the same in *tracheal* as in laryngeal stenosis. The differential points are: —
In laryngeal stenosis the larynx makes effectual respiratory movements (it goes upward in inspiration, downward in expiration); in tracheal stenosis it remains quiet or almost so. The voice is usually free in tracheal stenosis; in laryngeal stenosis it is very hoarse or there is aphonia. In laryngeal stenosis the head is usually thrown back; in tracheal stenosis the chin approaches the chest. Concerning thrills, see below.

3. **Pain** in the larynx is frequently complained of and is described as tickling, pressure, burning, or soreness; inflammatory processes, particularly, produce severe pains, which radiate toward the ear. Pain does not furnish any especial diagnostic points.

4. **Cough.** — A *laryngeal cough* is sometimes of a particularly loud and barking character, and can frequently not be distinguished from a cough emanating from the lungs. It is called the *croup cough*, although by no means characteristic of croup, since pseudo-croup and other affections are responsible for coughs of similar sound. Cough is most easily produced from the posterior wall of the larynx (intra-arytenoid region). Irritation of the ventricular bands and of the vocal cords does not evoke a cough so easily.

5. **Expectoration** is common to most diseases of the larynx, but does not furnish any characteristic data as to its origin in the larynx.

6. **Dysphagia** is always a sign of severe and advanced disease of the larynx, and is especially characteristic of the involvement of the epiglottis and the posterior wall. The laryngeal dyspnœa and the change in the voice prevent diagnostic confusion with difficulties in swallowing caused by œsophageal stenosis. The changes in the throat which the dysphagia produces are visible.

The External Examination of the Larynx (Inspection and Palpation)

This is not of great importance. The visible respiratory movements of the larynx and their diagnostic significance in stenosis have been discussed above.

When stridor is present, one can feel a *thrill* in that part of the neck occupied by the trachea, which can be appreciated alone or stronger on expiration, when the obstruction to respiration lies deep in the trachea. In laryngeal stenosis, on the other hand, the thrill is palpable exclusively, or in the vast majority of instances, on inspiration.

Visible or even palpable pulsation of the trachea, or even of the entire larynx, has been described as a sign of aneurysm of the aorta.

The vibrations of the vocal cords in speaking may be felt equally on both sides by placing the index fingers at the sides of the thyreoid cartilages. Weakness of the vibrations of one side speaks for disease of that side; in combination with other symptoms (*e.g.*, a threefold division of the voice, see above), this sign may be the guide for a far-reaching diagnosis without the aid of the laryngoscope; the diagnosis of disease of the larynx is usually accomplished, however, by

Laryngoscopic Examination

The sunlight or the light of a lamp is thrown upon the laryngeal mirror, placed obliquely against the uvula, by a head mirror or reflector. The laryngeal mirror then gives the picture of the larynx.

The normal picture in the laryngeal mirror shows above (anteriorly) the epiglottis, below the posterior wall of the larynx, the intra-arytenoid region, the two arytenoid cartilages, and above these, as a slight prominence, the cartilages of Santorini and Wrisberg (cunicula laryngis and cuneiform cartilage, respectively). At the sides, the picture is bounded by folds of mucous membrane, which extend from the epiglottis (above) to the arytenoid cartilages (ary-epiglottic ligaments); the middle of the picture is taken up by the true vocal cords, which pass antero-posteriorly (from above downwards) and are divided into two parts, the ligamentous, the anterior two-thirds (formed by the arytenoid cartilages), and the cartilaginous, occupying the posterior third. The cleft between the cords is known as the rima glottidis, the anterior portion of which is called the vocal glottis, the posterior the respiratory glottis. The false vocal cords (ventricular bands) run parallel to and above the true cords, and are seen at the side in the picture. Between the two, lies the ventricle of Morgagni. The right vocal cord appears on the right side in the mirror, the left on the left side; there is no transposition; but the right vocal cord is, of course, to the left of the observer.

The laryngoscopic examination is not intended to estimate merely the normal appearance of the larynx, but above all the mobility of the vocal cords also; the patient must alternately phonate (say *ah*) and take deep inspirations. In phonation the epiglottis is raised and the view into the larynx facilitated. During inspiration the glottis is opened, the vocal cords are abducted; in phonation they are adducted, the glottis is closed.

The direct inspection of the larynx without a laryngoscope (*Kirstein's autoscopy*) is rendered possible by the use of the autoscope when the uppermost air-passages are rendered straight by a peculiar position of the head, when the root of the tongue is pressed forward, and when there is a simultaneous elevation of the glottis.

Normal Functions of the Muscles and Nerves of the Larynx

The opening of the glottis is accomplished by the crico-arytenoid muscle, innervated by the recurrent laryngeal nerve; it acts as an abductor by drawing the vocal process of the arytenoid cartilage outward.

Closing of the glottis. The intra-arytenoid muscle (*arytænoideus transversus* and *obliquus*) closes the glottis by drawing the arytenoid cartilages together; the other closers of the glottis are the crico-arytenoid lateralis, the principal adductor, which turns the vocal process inward, and thyreo-arytenoid internus, whose course is in the vocal cords. These muscles are innervated by the recurrent nerve, except the *arytænoideus transversus*, which receives a few fibres from the superior laryngeal nerve.

Tension and elongation of the vocal cords are accomplished by the crico-thyreoid muscle, the nerve supply of which comes from the external branch of the superior laryngeal nerve and which draws the thyreoid cartilage forward and downward toward the previously fixed cricoid cartilage.

Tension and shortening of the vocal cords: the thyreo-arytenoid internus muscle, supplied by the recurrent nerve.

There are two laryngeal nerves, therefore, the superior laryngeal, the inferior laryngeal, or recurrent, which are branches of the accessory nerve, and which contain motor and sensory fibres.

The *superior laryngeal nerve* supplies with its small external branch the crico-thyreoid muscle; the large inner branch perforates the hyo-thyreoid membrane and supplies the mucous membrane with sensory fibres. Its motor fibres are few in number, and go to the *arytænoideus transversus* and the muscles of the epiglottis (the *thyreo-epiglotticus*, which raises the epiglottis, and the *ary-epiglotticus*, which lowers it).

The *inferior laryngeal nerve* passes backward in the thoracic cavity, on the right side encircling the subclavian artery, on the left the arch of the aorta, passes to the larynx between œsophagus and trachea, and innervates with its internal branch the *crico-arytænoideus posticus* and the *arytænoideus transversus* muscles, and with its external branch all the remaining muscles of the larynx.

Symptoms of the Most Important Diseases of the Larynx

Acute laryngitis. — Usually no fever (in children, irregular remittent fever). Hoarseness and pain in the throat; slight difficulty in swallowing. Mucous or muco-purulent expectoration, not characteristic. *Laryngoscopic examination:* redness and swelling of the mucous membrane, diffuse or circumscribed. Vocal cords reddened, occasionally superficial erosions. Vocal cords apparently small in consequence of the swelling of the ventricular bands.

Chronic laryngitis. — Pressure, scratching, etc., in the larynx, especially in singing, smoking, etc.; speech thick, hoarse; aphonia after continued speaking. Scanty, muco-purulent sputum. *Laryngoscopic examination:* dirty gray-red discoloration of the vocal cords, profuse secretion, occasional swelling of the follicles of the mucous membrane to the size of a grain of sand (*laryngitis granulosa* or *follicularis*). Very often, callous thickening, grayish-red in color, of the posterior wall of the larynx.

Tuberculosis of the larynx.— Usually secondary to pulmonary phthisis. Hoarseness, aphonia, pain, dysphagia. *Laryngoscopic examination,*— 1st stage: *tubercular infiltration.* Mucous membrane, puffy, swollen, pale, œdematous. 2d stage: *tubercular ulceration.* Irregular, confluent swellings with jagged, raised edges, passing into the depths. The pit of the ulcer covered with purulent secretion in which tubercle bacilli are present. Sites of predilection for tubercular changes are, first, the arytenoid cartilages and the intra-arytenoid space, next, the true and false vocal cords.

Syphilis of the larynx.— The complaints and subjective symptoms are usually not characteristic. *Laryngoscopically* one may differentiate. 1. *Early forms* (secondary phenomena): erythema of the larynx, sharply limited red spots; syphilitic laryngitis, usually not to be distinguished from non-specific chronic inflammation (history and examination for other syphilitic lesions, especially in the throat); *papules*, grayish-white, flat elevations on the inflamed mucous membrane (mucous patches); superficial ulcerations (sites of predilection: epiglottis, especially its border and lingual surface, the vocal cords). 2. *Late forms* (tertiary phenomena): circumscribed gummata and diffuse gummatous infiltrations characterized by rapid necrosis; the ulcers following the necrosis are deep and are sharply limited, but can not always be easily distinguished from tubercular ulcers. The differential diagnosis is established by the demonstration of an existing pulmonary phthisis (tubercle bacilli in the sputum), or of a previous syphilitic infection; in doubtful cases by *the result of antisyphilitic treatment.* Luetic ulcers may become very important by the strong contraction of radiating scars, which may be the source of very severe laryngeal stenosis.

Tumors of the larynx.— Usually involving the true or false vocal cords, pediculated (polyps) or sessile. Symptoms dependent upon the size and situation of the growth; there is usually hoarseness, often the sensation of a foreign body, pain, dyspnœa, sometimes dysphagia. It is essential to differentiate between benign and malignant growths (carcinoma, rarely sarcoma). The benign tumors, frequently with a smooth surface, are always circumscribed; carcinoma, on the contrary, usually not sharply limited, of irregular form, soon ulcerating; swelling of the lymph-glands demonstrable. The motility of the vocal cords soon diminished in the presence of a malignant growth. Age and condition of patient's strength important. Diagnosis often difficult, especially in the early stages, in which it is important on account of the therapeutic measures. Usually a positive diagnosis can be made only by the microscopic examination of a small piece of the tumor secured by excision. The special anatomical diagnosis of *benign* tumors (fibroma, papilloma, myoma, adenoma, etc.) is only to be made by the microscopic examination after extirpation or the removal of a small piece by excision.

Pachydermia laryngis are wart-like thickenings on the most posterior parts of the vocal cords, usually containing a depression on the surface, occasionally with gaps or clefts, and then not unlike carcinoma; usually accompanies chronic catarrh.

Spasm of the glottis (laryngo-spasm): spasmodic closure of the glottis leading to the point of suffocation and unconsciousness. In children, stands in etiological relation to rachitis; in adults, usually caused by hysteria, but may be due to organic disease of the central nervous system (laryngeal crises of locomotor ataxia).

Paralysis of the Vocal Cords

When of peripheral origin, paralysis of the vocal cords may be produced by pressure upon the recurrent nerve (by goître, aneurysm of the aorta, carcinoma of the œsophagus, enlarged lymph-glands, tumors of the mediastinum, pericarditis), or in consequence of a neuritis (alcoholic or rheumatic); of *central* origin: hysteria (*functional* paralyses of the larynx), or as a result of *organic* disease of the nucleus of the accessory nerve in the floor of the 4th ventricle (in locomotor ataxia, multiple sclerosis, bulbar paralysis, syringomyelia, cerebral syphilis, etc.).

Differentiation: 1. **Paralysis of the abductors of the vocal cords.** — Paralysis of the external abductors (crico-thyreoid muscle) recognized by palpation of the space between the thyreoid and cricoid cartilages; the normal approach of the two in phonation is absent. The glottis is not tense; it appears as a wavy line; voice hoarse.

This is a paralysis of the superior laryngeal nerve (most frequently a post-diphtheritic paralysis), in which the depressors of the epiglottis do not act (epiglottis immovable, usually directed toward the root of the tongue), and the sensibility of the mucous membrane of the larynx is lost. These three factors (elevation of the epi-

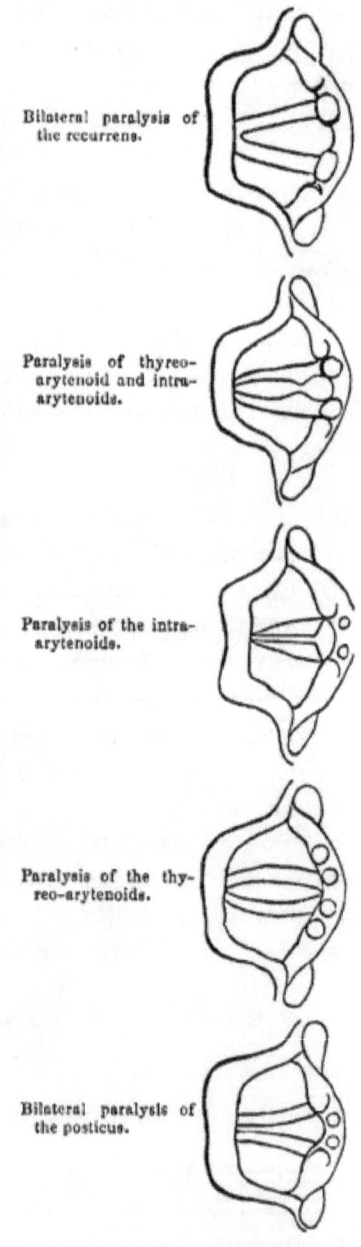

FIG. 25.

glottis, anæsthesia of the mucous membrane, atony of the glottis) are responsible for the dangers of mistakes in swallowing (foreign-body pneumonia).

2. **Paralysis of the adductors of the glottis.** — The paralysis of the adductors is usually functional. The glottis stands widely open (inspiratory position through action of the antagonistic muscles), when all three adductors are attacked (*crico-arytænoideus lateralis, thyreo-arytænoideus internus*, and *arytænoideus transversus* — complete paralysis of the adductors). The glottis then has the form of a triangle, in consequence of which there is no interference with respiration, but complete aphonia exists.

Isolated paralysis of the arytænoideus transversus muscle (frequent in acute catarrh and hysteria) shows the cartilaginous portion of the glottis gaping as a small triangle while the anterior part of the glottis closes in phonation. Hoarseness, even aphonia, without dyspnœa, supervenes.

Exclusive paralysis of the crico-arytænoideus lateralis is very rare and scarcely to be diagnosticated positively. On phonation a small cleft of the glottis near the vocal processes remains open; the disturbance of the voice is meagre.

Paralysis of the *thyreo-arytænoidei interni* is the most frequent; on phonation a small oval of the glottis, especially in its centre, remains open; the edges of the poorly vibrating, narrow vocal cords appear *excavated;* the voice is *aphonic;* no dyspnœa. This type of paralysis is common as an *occupational neurosis* (singers, public speakers, etc.), and usually presents the picture of *hysterical aphonia* (rarely total paralysis of the adductors).

Functional hysterical aphonia may be recognized by its sudden appearance and disappearance (intermittent aphonia is almost always hysterical). It may accompany psychical disorders; the cough usually has a ringing tone. Mere pressure upon the thyroid cartilages produces the peculiar tone of the cough. The history and the other symptoms render the diagnosis that of hysteria.

3. **Paralysis of the abductors of the glottis** (*posticus paralysis*). — The paralysis of these muscles is always organic, never functional. In paralysis of the *crico-arytænoideus posticus* muscle, the vocal cord lies in the median line, through the action of antagonistic muscles; phonation is therefore more or less intact.

Unilateral abductor paralysis: no dyspnœa; voice sometimes rattling and not clear, more frequently intact; no symptoms then arise, but the paralysis is accidentally discovered (in syringomyelia, tabes dorsalis, etc.).

Bilateral abductor paralysis: both vocal cords immovable, lying in or very near the median line (stenosis of the glottis); voice almost normal, but severe inspiratory dyspnœa with stridor. Expiration easy. The closely lying vocal cords approach each other more closely in inspiration. Tracheotomy necessary.

4. **Paralysis of the recurrent laryngeal nerve.** — If all the branches of the recurrent nerve are paralyzed, adduction and abduction stopped, the vocal cords take a position between the two: *cadaveric position*.

Unilateral paralysis of the recurrent nerve. The paralyzed vocal cord in cadaveric position with excavated edge; the arytenoid cartilage dropped forward toward the interior of the larynx; the unaffected cord often passes beyond the median line in phonation, so that a closure of the oblique glottis supervenes (over-riding of the arytenoid cartilages); voice weak, with no ring; no dyspnœa.

Bilateral paralysis of the recurrent nerve. Both vocal cords in cadaveric position, glottis widely opened in the same position in phonation and respiration. *Absolute aphonia* (cough and expectoration weak, and with little sound); no dyspnœa.

In the recurrent nerve, the adductor and abductor filaments lie separated. The abducting nerves are physiologically more sensitive than the adductors. All influences which lead to recurrent paralysis (gôitres, aneurysms, etc., see above) frequently cause a primary abductor paralysis which after a longer or shorter time (sometimes years) passes into a recurrent paralysis as soon as the adductor part of the nerve suffers from the disease.

CHAPTER VI

THE DIAGNOSIS OF DISEASES OF THE RESPIRATORY TRACT

In the history of the patient, particular stress must be laid upon *hereditary taint* when tuberculosis is suspected. The so-called *scrofulous* affections of childhood and previous pulmonary diseases are important, as are *hæmatemeses* and inflammations of the bones or joints of a probable tubercular origin. The occupation (stone-cutters, coal-miners, compositors, etc.) must also be taken into account, and sometimes a weakening condition, such as a recent puerperium, may offer a basis for pulmonary invasion. When lung trouble of gradual origin is before the physician, the early symptoms of phthisis pulmonalis must be looked for.

The attention of the physician is directed toward the respiratory apparatus on complaint of *pain in the side of the chest, cough,* and *expectoration* or *dyspnœa*.

Pain in the side is caused mainly by affections of the pleura, and calls for an examination of the thorax. In differential diagnosis, it is not particularly valuable, since a mild pleurisy may accompany many other thoracic diseases.

Although *pain in the chest* may be called forth by some cardiac affection or by a muscular rheumatism, it demands a careful diagnostic examination. It may occur in any part of the chest, and may be described by the patient as a pressing or drawing pain.

A *cough* is almost invariably a sign of disease of the respiratory apparatus. It is reflex in its nature, and is caused by an irritation of the mucous membrane of the larynx, trachea, or bronchi by an accumulation of secretion. It can not arise from the alveoli of the lungs, although irritation or inflammation of the pleura may produce it. Occasionally a cough may be induced by an irritation of the pharynx, œsophagus, stomach, liver, or spleen.

The *kind* of cough may, at times, be of diagnostic significance.

It is important to notice the *number* of coughs that succeed each other in each attack. The so-called *hacking cough* is common in incipient phthisis, and may direct the physician's attention to the latent disease. The cough of *pertussis* is characteristic, and is seldom heard in other pulmonary diseases. It consists of a great number of coughs rapidly following each other and interrupted by deep, sighing, stridulous inspirations. A *dry* or *moist* cough, a *hard* or *loose* cough, give indications as to the amount and character of secretion and as to the progress of the disease. The *pitch* and *noise* made by a coughing patient will sometimes allow an estimate to be made of his general condition. A particularly loud, barking cough, for instance, is characteristic of inflammation of the larynx and trachea. A differential diagnosis, except in pertussis, can rarely be made from the mere symptom, cough; but the intensity of a pulmonary lesion, or its progress, may be recognized by the cough. Cardiac diseases sometimes engender a cough after a secondary pulmonary congestion has arisen.

Of the greatest importance in the diagnosis of diseases of the lungs is the *expectoration* or *sputum*. It must not be forgotten, however, that a catarrh of the nose or throat may be responsible for expectorated material, brought forth by *hawking*. Sputum which is expectorated by a cough comes from the larynx, the trachea, the bronchi, or from an ulcerating process in the lungs. Although the examination of the sputum usually concludes the examination of the respiratory organs, it is as well to take note of pathological peculiarities at once; as, for instance, bloody sputum or sputum of rusty-red or green color; balls of sputum or expectoration that sinks in water, etc.

Inspection of the Thorax

The inspection of the thorax gives information: 1. As to the *normal shape* or *deformity of the chest*.

Curvatures of the spine have a diagnostic importance, since they are etiological factors in some diseases of the lungs: through the compression of parts of the lungs and the compensatory emphysema of other parts; in the course of time this may be followed by dilatation of the right ventricle, and cyanosis and dyspnœa in consequence of the diminished circle of the pulmonary circulation.

Kyphosis is the term applied to a posterior spinal curvature, and is known as *gibbus* when it forms an acute angle. A forward curvature is known as *lordosis*, while a lateral curvature is called *scoliosis*. The most common form is the combination of kyphosis and scoliosis; *i.e.*, a simultaneous lateral and posterior curvature.

Anomalies of the sternum have little diagnostic importance, as they do not produce pulmonary lesions. The xiphoid process may be pressed inward in the persons of men who, in their occupation, hold tools or implements against it, as shoemakers. A deep sinking in of the lower part of the sternum gives rise to the *funnel-shaped chest*: it is a congenital deformity, and is usually associated with other physical or mental abnormalities. A *pigeon* breast or *wedge-shaped* breast (*pectus carinatum*) is caused by rachitis by the compression of the costal cartilages. In this manner the sternum is pushed forward and assumes the shape of a wedge.

The angular eminence seen in many individuals between the manubrium and the body of the sternum is described as the *angle of Ludwig* (*angulus Ludovici*). The average length of the sternum in the adult is from 16 to 20 cm.

2. Of great diagnostic value is the circumference of the chest, — normal, *diminished or enlarged;* and it is important to note whether the enlargement or diminution exists on *one* or *both* sides.

a. A *bilateral increase* in the circumference of the thorax presents a characteristic picture; the chest is very broad, but seems shortened. It is known as the *barrel-shaped chest*, — and is pathognomonic for emphysema (*volumen pulmonum auctum*). Although the enlargement includes all the thoracic diameters, the antero-posterior is mainly deepened. The thorax is in a position of permanent inspiration.

The entrance of air or fluid into the pleural cavity of either side may produce an *enlargement of that half* of the thorax. The affected side *usually* lags behind in the respiratory movements. An unilateral enlargement occurs in pneumothorax and pleurisy with effusion, less often accompanying mediastinal tumors.

b. A *bilateral retraction* of the chest-walls is usually congenital. The thorax is long, flat, and narrow, the antero-posterior diameter is diminished, and the intercostal spaces are broad. This form of thorax has been described as a *paralytic thorax*, and generally arouses the suspicion of phthisis.

An *unilateral retraction* of the thoracic parietes, or better, perhaps, a sinking in of one side, is found after the absorption of pleuritic exudates and in the shrinking of tubercular lungs (*cirrhosis pulmonum*).

Measurements of the thorax. Dilatation or narrowing can usually be estimated quite distinctly with the naked eye; but it is sometimes necessary, in order to establish the circumference or measurements of the chest, to employ a tape-measure or a cyrtometer.

The mean average measurements of the thoracic circumference in a healthy adult man, estimated at the point of greatest expansion, are 82 cm. after deepest expiration and 90 cm. after deepest inspiration. This is, of course, only approximate, and varies largely in individuals. In general, it may be said that the circumference of the chest at the nipple after deepest expiration should equal one-half, at least, of the body's length. The right half of the chest may be from 0.5 to 2 cm. larger in a right-handed person than the left side, and *vice versa*. The thoracic diameters in women are somewhat less than the corresponding ones in men.

The sterno-vertebral diameter, measured with the cyrtometer, averages in adult men superiorly 16.5, inferiorly 19.2 cm. The lateral diameter at the level of the nipple (costal diameter), 26 cm. In women all these diameters are diminished.

3. The *frequency*, type, and rhythm of the *respiratory movements* may be appreciated by inspection.

The normal number of respirations varies from 16 to 20; in the new-born from 40 to 44. The normal relation between pulse and respiration is 1 to 3.5 or 1 to 4.

The lungs remain completely passive in respiration, merely following the motion of the ribs and diaphragm. In men the inspiratory expansion is due largely to the contraction of the diaphragm, and is known as the *costo-abdominal type* of breathing. The *costal type* is found in women in whom inspiration is principally accomplished by the elevation of the ribs by the intercostal and scaleni muscles.

An increased frequency and deepening of respiration (*dyspnœa*) is often caused by cardiac diseases and extensive abdominal distention.

Dyspnœa is to be referred to some pulmonary lesion when there is simultaneous cough and expectoration or pain in the side or other parts of the chest. It is a symptom of pneumonia, when accompanied by high fever and rubiginous sputum; it occurs in pleurisy and pneumothorax with impaired mobility and expansion of the affected side; it is seen in emphysema in connection with a barrel-shaped chest and diffuse bronchial râles over both lungs; it is observed, finally, in advanced phthisis (*habitus paralyticus*).

Dyspnœa which appears in a tubercular subject in an early stage of the disease is significant of some complication which must be recognized; as, empyema, pleurisy, pneumothorax, or miliary tuberculosis.

The kinds of dyspnœa. Inspiratory and expiratory dyspnœa are recognized. In inspiratory dyspnœa the accessory muscles of inspiration (the sterno-cleido-mastoid, scaleni, levatores costarum. pectorales major and minor, levatores scapulæ, rhomboidei. trapezii, and erectores trunci) are tetanically contracted. In extreme cases there is a retraction of the lower ribs and of the space just below the ensiform appendix. Stenosis of the larynx, trachea. or bronchi may call forth a high grade of inspiratory dyspnœa.

In expiratory dyspnœa, expiration is prolonged and difficult. The abdominal muscles, the quadratus lumborum, and the serratus posticus inferior act as accessory muscles. This variety of dyspnœa is found principally in emphysema and bronchial asthma.

The dyspnœa most commonly seen is a mixture of expiratory and inspiratory "air-hunger."

Asthma is that form of difficult breathing in which dyspnœa appears in attacks lasting at times for hours and followed by a longer or shorter period of quiet, natural respiration. Its most common form is *bronchial asthma*, probably of neurotic origin, which produces in apparently healthy individuals a temporary expiratory dyspnœa of the most intense kind. During such an attack the diaphragm is tetanically contracted, the liver dulness is lower than normal, and diffuse, sibilant râles may be heard. At the end of the attack a small quantity of characteristic sputum is expectorated (see below). The pulse remains regular and strong during the attack. The prognosis after a single seizure is good so far as life is concerned; oft-repeated onsets lead to the development of emphysema.

In striking contrast to bronchial asthma are the dyspnœic attacks in cardiac diseases (*cardiac* asthma). In these cases the pulse becomes small and irregular and the left ventricle is dilated. The prognosis must always be doubtful.

Nasal asthma is spoken of when attacks of bronchial asthma are evoked by reflex irritation of a pathological nasal mucous membrane (p. 108). *Dyspeptic* asthma is the name given to attacks of fear and anxiety in persons suffering from nervous dyspepsia. (In these instances one has usually to do with a neurasthenic.) *Urœmic* asthma consists of the attacks of dyspnœa occurring in chronic nephritis, which are probably in reality cardiac asthma. *Hay-fever* produces attacks of dyspnœa in otherwise healthy persons. It is probably caused by irritation of the nasal mucous membrane by the pollen of certain grasses.

Spirometry

By inspection one may conclude as to the spirometry; that is, the determination of the quantity of air given off by the deepest possible expiration after the deepest possible inspiration (*vital capacity*). The vital capacity is *diminished* in all diseases of the respiratory organs.

The diagnostic value of this measure is unimportant, since there are no characteristic differences in the vital capacity among the various lung diseases.

On the other hand, however, spirometry is of real value in judging of the improvement or aggravation of a disease, particularly with reference to the influence of the therapeutic measures involved.

The vital capacity is measured by Hutchinson's spirometer. In healthy men it varies from 3000 to 4000 c.c., an average of 3600 c.c.; in healthy women it lies between 2000 and 3000, an average of 2500 c.c. The vital capacity increases with the bodily length, each centimetre of the latter representing about 22 c.c. of the former. In children and in the aged it is diminished.

Complementary air is that which may be drawn into the lungs by the deepest inspiration after an ordinary inspiration. It is about 1500 c.c.

Reserve air is that which may be expired by the deepest expiratory effort after ordinary expiration. It equals about 1500 c.c.

Breathing or *tidal air* is the air ordinarily changed by each act of respiration. It amounts to about 500 c.c.

Residual air is the air remaining in the lungs after the most violent expiratory effort. It is estimated at from 1600 to 2000 c.c.

After the thorough consideration of the conclusions reached by the inspection of the thorax and its movements, the *physical examination* of the respiratory organs is undertaken: the *percussion* and *auscultation* of the thorax.

MENSURATION

The topographical data which enhance the ease of the determination of the height and breadth of the chest are here given.

The *height* of the chest is *anteriorly* determined by the clavicle or by the supra- and infra-clavicular spaces. The external border of the latter includes the space of *Mohrenheim*. Below the clavicle the ribs are employed. The 2d rib is the starting-point in counting, since it can be easily recognized by its attachment to the sternum (angle of Ludwig). The nipple usually lies over the 4th rib or in the 4th intercostal space, usually about 10 cm. distant from the edge of the sternum. At the level of the xiphoid or ensiform process a distinct furrow usually traverses the thorax, which marks the site of insertion of the diaphragm (*Harrison's furrow*). The region beneath this furrow to the arch of the ribs is the *hypochondrium*. The determination of the height of the chest *posteriorly* is aided by using the spine of the scapula as a landmark; by this means the 7th cervical vertebra (*vertebra prominens*) is easily felt. The other guiding-point is the scapula, which covers the space from the 2d to the 7th or from the 3d to the 8th ribs and which is divided by its spine into the supra- and infra-spinous fossæ. The space between the internal borders of the scapulæ is the inter-scapular space.

For the determination of the *width* of the thorax, the following imaginary lines, drawn parallel to the long axis of the body, are used: —

1. The *sternal line*, representing the borders of the sternum or the attachments of the ribs.
2. The *para-sternal line*, midway between the sternal line and the nipple.
3. The *mamillary line*, drawn through the nipple.
4. The *anterior axillary line*, drawn through the anterior border of the axilla (pectoralis major).
5. The *middle axillary line* is drawn through the middle of the axilla.
6. The *posterior axillary line* is drawn through the posterior border of the axilla (latissimus dorsi).
7. The *scapular line*, drawn through the inferior angle of the scapula.

The linea costo-articularis is drawn from the claviculo-sternal articulation to the apex of the 11th rib.

Topography of the individual lobes of the lungs. The right lung has 3 lobes, the left but 2. The *right upper and right lower lobes* may be made out *posteriorly on the right side*. The border between them begins at the level of the 2d or 3d rib. This border is divided about 6 cm. above the angle of the scapula into two fissures which enclose the middle lobe. The superior fissure is directed almost horizontally forward and reaches the anterior border of the lung at the level of the 4th and 5th costal cartilages. The inferior fissure is directed perpendicularly downward and reaches the inferior border of the lung in the mamillary line. Hence we have: *right side, anteriorly, upper lobe* as far as the 3d rib; below that, *middle lobe*. On the *left side* posteriorly, upper lobe as on the right side. The border extends without bifurcation obliquely forward and ends in the mamillary line at the 6th rib. Hence we have: *left side, posteriorly, upper and lower lobes; anteriorly only the upper lobe* (and the heart).

Percussion of the Thorax[1]

By the percussion of different areas over the thorax, characteristic differences in sound may be appreciated according as the thoracic organs contain more or less air.

Percussion serves to determine: —

1. The borders of the air-containing lungs as differentiated from other organs (as the liver, the heart).
2. The amount of air contained in the lungs, which differs characteristically in diseases of these organs.

The *qualities of sound* obtained by percussion are: —

1. Clear or dull = loud and low.
2. High and low (pitch).
3. Tympanitic and non-tympanitic.

One obtains a clear (loud) non-tympanitic note over the lungs; a clear (loud) tympanitic note over gastric and intestinal areas; a dull note over the heart, the liver, the spleen. *High and low* are qualities referred only to tympanitic sounds (as over cavities).

Characteristic qualities of sound are: the *metallic sound* and the *cracked-pot sound* (*bruit de pot fêlé*).

It is essential for the beginner to practise percussion frequently on healthy persons, in order to drill his ear to the different qualities of the clear (normal) notes obtained over the lung. The normal note differs in intensity over the same lung, depending upon the muscular thickness and the deposit of fat. It is important to accustom one's self to compare the note obtained over analogous areas of the two halves of the body.

The Percutory Determination of the Borders of the Lungs (Fig. 26)

Upper border: the determination of the upper borders of the lungs is important; because unilateral depression of the

[1] Percussion was discovered by Auenbrugger in Graz (1722–1809).

upper border is frequently the first sign of pulmonary tuberculosis. In healthy persons, the upper border extends from 3 to 4 cm. over the upper border of the clavicle; posteriorly it lies at the spinous process of the 7th vertebra. In healthy individuals, both borders are at the *same* level.

As the limit of the upper border is fixed, it may be simultaneously determined if the percussion note on one side is less clear than on the other. Differences in sound at the apices, in most instances point to tuberculosis (see below).

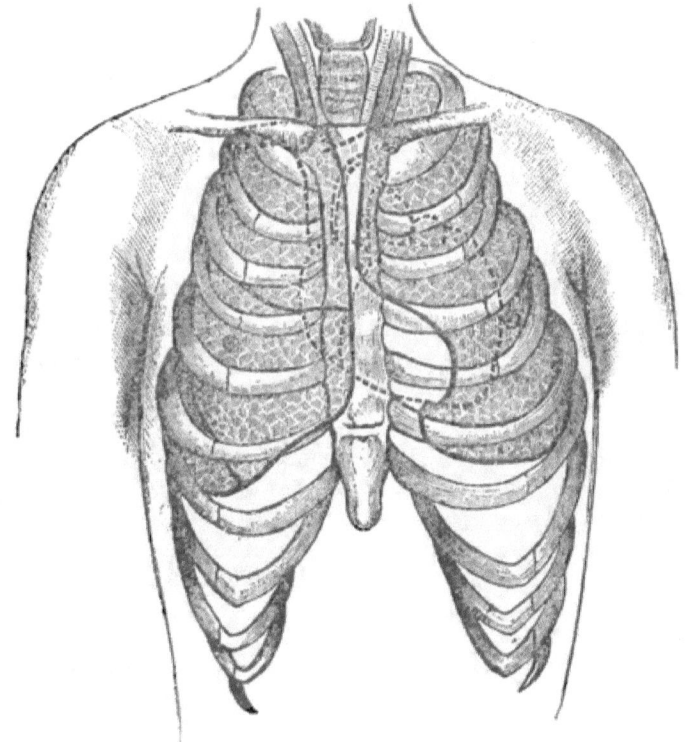

Fig. 26. — Relative Positions of Thoracic Viscera.

Lower border: while the determination of the upper borders of the lungs furnishes data for the diagnosis of phthisis, the establishment of the lower border aids in the diagnosis of emphysema (*volumen pulmonum auctum*). In emphysema, the lower borders of the lungs are *lower* than in health.

The lower border of the lungs lies:—

On the *right* side, at the sternal border on the 6th rib; in the mamillary line, at the lower edge of the 6th or the upper border of the 7th rib; in the anterior axillary line, at the lower border of the 7th rib; in the scapular line, at the 9th rib; next to the vertebral column at the spinous process of the 11th dorsal vertebra.

On the *left* side, the lower border of the lung is difficult to make out, on account of the proximity of the stomach, the tympanitic sound of which merges gradually into the non-tympanitic sound of the lung.

It is customary to determine the lower border of the lung in the *right mamillary line*. In healthy persons, the relative liver dulness begins at the lower border of the 4th rib. The absolute liver dulness, *i.e.*, the lower border of the lung, begins at the lower border of the 6th or the upper border of the 7th rib. If one obtains a long clear note of some duration from the 7th to the 9th ribs, the diagnosis of emphysema is assured.

Temporary lowering of the lower border of the lung is found in attacks of bronchial asthma. The lower border of the lung is raised in meteorism, ascites, abdominal tumors; on one side only, indicates contracting pleurisy or atrophy of pulmonary tissue. The lower border of the lung on the right side is also pushed upward in pleurisy with effusion. This fact is demonstrated, of course, posteriorly; inferiorly on the right side.

Respiratory changes in the lower borders. In quiet, normal breathing, the lower borders sink about 1 cm. in the mamillary line; from 3 to 4 cm. in deep inspiration. The upper borders, in quiet respiration move upward about ½, in deep inspiration about 1½ cm. The respiratory displacement depends upon the position of the body. In the dorsal position, the anterior lower border of the lung moves 2 cm. lower than in the vertical position. The displacement in the axillary line is most marked, and may reach 10 cm. in deep inspiration with the patient in the side position.

Respiratory displacement is absent in extensive adhesions between the pulmonary and costal pleuræ, and is much diminished in emphysema.

Dulness over Pulmonary Areas

Dulness over the normal extent of the lungs may arise:—

1. When the lung *immediately* below the chest-wall is devoid of air; this area must have an extent of at least 4 cm. Pulmonary tissue may become deprived of air by *infiltration* and by *atelectasis*.

 a. Infiltration is due to pneumonia, tuberculosis, abscess, more rarely gangrene, hæmorrhagic infarct, new growths, aneurysm.

 b. Atelectasis arises from compression (pleuritic or pericarditic exudate and new growths) or from the resorption of air when the bronchi become occluded (by tenacious secretions or tumors and by syphilitic stenosis).

2. When *fluid* is exuded into the pleura, between the lung and the chest-wall (pleuritic exudate, hydrothorax); the quantity must be at least 400 c.c. Dulness is evoked, also, by thickening of adherent layers of pleura (*adhesions*).

Dulness over the upper lobes (apex) (with good resonance behind, below) usually indicates pulmonary phthisis, more rarely pneumonia, gangrene, or new growths. Dulness over the lower lobes (posteriorly, below) usually indicates pneumonia or pleurisy. The differential diagnosis is made by auscultation. Tubercular infiltration of the lower lobes is usually secondary to extensive tubercular invasion of the upper lobes. In rare cases, dulness of the lower lobes is to be referred to gangrene, infarct, or tumors. In bronchitis and miliary tuberculosis, there is *no* dulness present.

In protracted diseases, with the patient constantly in the dorsal position, serous infiltrations causing dulness may appear in the lower portions of both lungs (*hypostatic congestion*).

TYMPANITIC RESONANCE IN THE AREA OF THE THORAX

In the healthy thorax, a loud tympanitic note is obtained only over the stomach (*semilunar* space, p. 83), and over the parts of the left lung immediately adjacent to it. These parts are thin enough to allow the tympanitic note of the stomach to be evoked by percussion.

The lower portion of liver dulness may be concealed by the tympanitic note due to tympanites.

A tympanitic percussion note in other parts of the thorax arises:—

1. In *cavities* in the parenchyma of the lung; the cavity must have at least the size of a large walnut, must lie near the chest-wall or must be connected with it by infiltrated tissue. Such cavities are present in phthisis, less often in bronchiectasis and gangrene.

2. In a collection of air in the pleural sac (*pneumothorax*), but only while the air is not under too great tension. Otherwise the note is deep and loud, but *not* tympanitic.

3. In complete infiltration of large pulmonary areas, by which there is a good conduction of sound between the broncho-tracheal column of air and the chest-wall; for instance, in tubercular infiltration of an entire upper lobe, and in the second stage of pneumonia (dulness with tympanitic sound).

4. In diminished tension (relaxation, atony) of the pulmonary tissue: over pleuritic and pericarditic exudate and pneumonic infiltration; frequently in the first and third stages of croupous pneumonia and in œdema of the lungs.

All conditions in which a tympanitic note is heard have this in common: that a column of air *above* is brought into vibration (without vibration of the tissues of the lung).

In the cases of cavities and pneumothorax, the column of air in the space is percussed directly; in the instances of extensive infiltrations, one percusses the normal cavities (broncho-tracheal column of air) through the thickened tissues. Pulmonary tissue lying above exudates is robbed of its elasticity and vibratory power, and hence, in percussion, the air only contained in the lung vibrates.

A **metallic** (amphoric note) differs from a tympanitic note by its higher pitch and by its longer duration. In addition to the fundamental note, there are overtones to be heard, which slowly disappear. They may be compared to a metallic echo.

Amphoric resonance may be elicited over the thorax : —

1. In the presence of cavities with homogeneous walls, which must have the size of a man's fist at least (6 cm. diameter).

2. In pneumothorax, when the air present is under a certain pressure, not too great.

In order to appreciate amphoric resonance more distinctly, percussion by means of the *hammer* and *pleximeter* may be employed (*mediate* percussion). While performing auscultation over the cavity, the pleximeter, placed next to the stethoscope, is lightly hit by the hammer. By this means a clear, metallic note is elicited.

The *cracked-pot sound* (*bruit de pot fêlé*) is brought out by a short, sharp percussion. It is best obtained with the patient's mouth open, over a superficial *cavity* which communicates with a bronchus by a narrow opening. The sound may resemble that of the *rattling of coins*. Caution must be observed in estimating the value of the cracked-pot sound, as it may appear in healthy persons, especially children, in speaking and singing; and moreover, is sometimes found in tissues relaxed and infiltrated (pleurisy and pneumonia).

Changes of Sound and Symptoms of Cavities

A tympanitic note is *high* or *low* according to the length of the column of air which produces it, and the breadth of the opening by which it reaches the external air. The shorter the column of air and the wider the opening, the higher the note. By having a patient open his mouth, the opening, communicating with a cavity which is freely connected with a large bronchus, may be enlarged; by changing the position of a patient, the diameters of a purulent cavity, with unequal diameters, may be changed. By thus changing the physical conditions, high and deep notes may be artificially produced. This is known as sound mutation or change.

1. The *change of sound* known as *Wintrich's* consists of the change produced in a tympanitic note by the opening and closing of the mouth — higher in the former, lower in the latter case. Found in lung *cavities* and *pneumothorax when these communicate freely with a bronchus.* Rarely in pneumonia and over pleuritic exudates.

One can imitate this phenomenon by percussing the larynx with alternate opening and closing of the mouth.

An *interrupted Wintrich's sound change* is the designation given to the disappearance of the phenomenon when the patient sits up, while it was easily obtained when he was lying down, and *vice versa*. It is caused by the occlusion of the bronchus, which in a certain position of the body freely opens into the cavity.

2. *Gerhardt's change of sound:* when the patient sits up, the tympanitic note is deeper than when he is lying. This phenomenon appears in oval *cavities* which are, in part, filled with fluid; the note has the lowest pitch when the longest diameter is horizontal. When this diameter is diminished by change of position, the note becomes clearer. (If the note becomes higher when the patient sits up, the diagnosis of a cavity is not certain.) This phenomenon may be produced by percussing a partly filled bottle.

3. *Respiratory change of sound:* in very deep inspirations, the tympanitic note over *cavities* sometimes rises in pitch, probably by the increased width of the glottis, which represents an increase in the size of the cavity.

4. *Biermer's change of sound:* percussion over a *pneumothorax* containing fluid, the percussion note is deeper in the recumbent than in the sitting posture, because the long diameter of the air-containing pleura becomes increased, in the lying position, at the expense of the long diameter of the exudate, which sinks from the diaphragm upon the posterior chest-wall.

Auscultation of the Thorax[1]

By means of auscultation one appreciates: (1) the respiratory murmur; (2) râles and friction sounds.

The Respiratory Murmur

Vesicular, bronchial, amphoric, and *broncho-vesicular* breathings are distinguished.

Vesicular breathing (*cell-breathing*) is found over the entire healthy lung. It is principally audible on inspiration; of a shuffling character, and is not heard on expiration, or if heard, is short and uncertain, and is rarely audible as a vesicular murmur.

Vesicular breathing may be easily imitated by placing the lips and teeth as though one were to pronounce a soft *f* and then taking a deep *inspiration*. Vesicular respiration arises in the trachea and the large bronchi, and is therefore really bronchial breathing, which attains its peculiar shuffling character through being conducted into the small bronchi and the alveoli of the lung.

Clear, soft vesicular breathing without râles is a positive sign that the auscultated area of lung is healthy.

Diminished intensity of vesicular respiration is found in emphysema, since, in consequence of the distention of the lung tissues, only a small quantity of air enters the alveoli; in pleurisy, because the fluid between the chest-wall and the lung is a poor conductor of sound; in instances of adherent and thickened pleuræ, because the lung enveloped by them can not unfold itself. Over very large pleuritic exudation, the respiratory murmur is entirely absent.

An *increased intensity* of *vesicular breathing* is normal in children: *puerile respiration*. It appears in swelling and stenosis of the bronchi and in bronchitis (because the current of air is inspired with greater force to overcome the obstructions present).

[1] Auscultation was discovered by the Parisian clinician Laennec (1781–1826).

An *increased intensity* of *vesicular expiration* and a prolongation of it arises when there are obstructions to the exit of air from the alveoli because of narrowing of the smaller bronchi, as in bronchitis and bronchial asthma. Prolonged expiration increased in intensity over *the apex of the lung* is an early sign of phthisis.

Interrupted or *jerky respiration* is vesicular breathing in which inspiration is accomplished with several intermissions. It may appear in healthy persons who breathe slowly and irregularly; but it is often found at the apex as an early sign of tuberculosis. It must not be given too high a value, and is to be considered only in connection with other symptoms.

Systolic vesicular breathing refers to the increased respiratory murmur often heard in the neighborhood of the heart during its systole. It has no diagnostic significance.

Bronchial breathing (*cavernous respiration*) is found in health over the larynx, the trachea, and the inter-scapular space. It is of a puffing character, principally audible on *expiration;* on inspiration it is usually heard shorter and weaker.

Bronchial breathing may be imitated by putting the mouth in the position of pronouncing the soft German *ch* and slowly *expiring*. Bronchial breathing arises through the rotatory motion imparted to the inspired air as it passes through the rima glottidis, which is in turn continued in the broncho-tracheal column of air.

Bronchial breathing is produced by analogous conditions to those which call forth a tympanitic percutory note:—

1. In large cavities, but the conducting bronchus must be free.

2. When the lung is so *consolidated* that the bronchial respiratory murmurs of the large bronchi are conducted unchanged to the chest-wall:

   (*a*) in infiltration by pneumonia or tuberculosis, less often gangrene;

   (*b*) in compression, particularly above exudates into the pleura.

Dulness and bronchial breathing over the lower lobe, with an uninvolved upper lobe, usually denotes pneumonic infiltration. Dulness and bronchial breathing over the apex, with a normal lower lobe, usually indicates tubercular infiltration. But it must not be forgotten that these symptoms may point to an upper-lobe pneumonia, a lower-lobe tuberculosis, gangrenous infiltration, pleuritic thickening, or compression due to tumors or aneurysms.

A loud, tympanitic note and bronchial breathing denote a cavity; a loud deep note with no respiratory murmur, a closed pneumothorax; dulness with no respiratory murmur, a pleuritic effusion.

Metamorphic respiration is the name given to an uncommon respiratory murmur which begins as vesicular and becomes bronchial. It is heard, when at all, over cavities, but is not at all characteristic.

Amphoric respiration is never heard over the healthy chest; it is a whistling respiratory sound with a metallic echo, produced under the same conditions as the metallic percussion note. It is pathognomonic for cavities with walls of uniform density of a diameter of at least 6 cm., and for an *open pneumothorax*. Over a closed pneumothorax no respiratory murmur is heard.

Amphoric respiration may be imitated by blowing over the top of a large bottle.

Broncho-vesicular breathing is breathing which is neither distinctly bronchial nor distinctly vesicular. One should be very cautious in diagnosticating this variety of respiratory murmur, and should try, by careful auscultation, and by having the patient take deep inspirations, to give the murmur a vesicular or a bronchial character.

Broncho-vesicular respiration is found in healthy persons, on superficial breathing, in the supra- and infra-spinous regions. In pathological conditions it is frequently heard without allowing any diagnostic conclusion to be drawn from it. When broncho-vesicular breathing is *permanently* heard at *one* apex, it may be regarded as a sign of beginning tuberculosis.

Râles and Friction Sounds

The adventitious respiratory sounds are always signs of a diseased condition of the mucous membrane of the lung, or of a collection of secretions, or of pus. Râles are known as *dry* and *moist*.

Dry râles are caused by the passage of the inspired air through a narrowed bronchus or one covered with a tenacious secretion. They may be of a *purring* or *sonorous* (*rhonchi sonori*), or of a *whistling* or *sibilant* (*rhonchi sibilantes*) character. They are pathognomonic for bronchitis.

Moist râles are heard when there is a collection of secretions which conduct the inspired air, or when bubbles containing air are ruptured, or when closed alveoli or bronchioles are forced open.

Moist râles may be *plentiful* or *scattered*, *fine* or *coarse*; they may be *ringing* or *not*, *metallic* or *not*.

The coarseness of râles depends upon the size of the cavity in which they arise; fine râles are usually derived from the small bronchi in a beginning infiltration; coarse râles in large bronchi and cavities.

Crepitant râles (bullous râles) take their origin in the forcing open of previously closed alveoli, and are a subdivision of fine râles. They may be imitated by rubbing the hair between the fingers. Crepitant râles are heard in the 1st and 3d stages of pneumonia, in œdema of the lungs, in miliary tuberculosis; sometimes, in the apices and the areas of lung immediately beneath them, in *healthy* organs during the first few inspirations after partial atelectasis.

Ringing râles are often heard under the same conditions as bronchial breathing (in cavities, in large infiltrations which act as sound conductors, in compression).

Metallic, ringing râles are often heard in connection with a metallic percussion note and amphoric breathing.

A single metallic râle is known as tinkling (*tintement métallique*, the sound of falling drops of water). Heard in pneumothorax.

Succussion (*succussio Hippocratis*) is a metallic splashing heard at some distance from a patient when the upper part of his body is energetically shaken from side to side. It is pathognomonic for the simultaneous presence of air and fluid in the pleural sac (sero- or pyo-pneumothorax).

Succussion may also be heard over a large stomach after the ingestion of fluids. It speaks in this instance in favor of gastric dilatation, but does not prove it.

Friction sounds (pleuritic) are heard when two adjacent layers of pleura, rubbing against each other in the respiratory act, become roughened by deposits of fibrin. They are pathognomonic for pleurisy, are most frequent in dry pleurisy, and in large effusions are heard usually in the stage of resorption. Friction sounds are absent in effusions with stasis (hydrothorax) and when adhesions between the layers of the pleura exist.

The distinction between friction sounds and dry râles is sometimes difficult to make. It may be remarked that râles are changed by coughing, friction sounds are not; that friction sounds are increased in intensity by the pressure of the stethoscope, and that under these circumstances a complaint of pain is made; that friction sounds are appreciable to the hand.

They may be compared to the creaking of new soles and to the rustling of a tense sail.

Pectoral Fremitus

Pectoral or vocal fremitus is appreciated by placing the hands symmetrically on the two sides of the chest while the patient speaks or counts in a deep voice. In healthy persons one feels a distinct vibration of the chest-wall produced by the conduction of the voice.

Pectoral fremitus is *increased* in pneumonia, above pleuritic effusions, in cavities with thickened walls. (Infiltrated and compressed tissues are good conductors of sound; in cavities the sound of the voice is intensified by reflection from the walls.)

Vocal fremitus is *diminished* or *absent:* —

a. When the pleural sac is filled with fluid or with air (*pleurisy* and pneumothorax). Over pleural adhesions, vocal fremitus is frequently well preserved.
b. When the large bronchi are occluded by tumors or stenosis.

The diminished intensity or absence of vocal fremitus is of diagnostic value only when the voice is strong and deep. A weak voice produces a weak fremitus. In very stout people, it can not be well felt.

Auscultation of the voice. The auscultation of a healthy chest during talking gives only an indistinct humming. This is *diminished* under the same conditions as vocal fremitus. The voice seems to be *intensified* when the waves of sound travel through good conductors, that is, through infiltrated or compressed tissues, or through cavities with thickened walls. Vocal auscultation is therefore bounded by the same limitations as vocal fremitus and bronchial respiration. An increased auscultatory intensity of the voice is called *bronchophony;* very much heightened intensity, *pectoriloquy.*

Ægophony is the name given to a peculiar sound of the voice, possessing a tremor and sounding like the bleating of a goat, which is often heard at the upper limit of an extensive pleuritic effusion.

Examination of the Sputum

The examination of expectorated material coughed up is indispensable for the diagnosis of diseases of the lungs. The examination begins with a simple inspection of the sputum (macroscopic examination) which is followed in case of necessity by a microscopic examination.

Every sputum may be placed in a group according to its main constituents: *mucous, purulent, serous, fibrinous, bloody* sputum; or it may represent a mixed form, as muco-purulent, bloody-serous, etc.

1. *Pure mucous sputum* (tenacious, glossy, sticking to the bottom of the glass) is characteristic of a beginning bronchitis. As bronchitis may be the forerunner of tuberculosis, a sputum consisting of pure mucus must be given cautious diagnostic judgment. The expectoration of the nose and pharynx is often purely mucous.

2. *Pure purulent sputum* (thick, confluent pus, not foamy) is found almost solely in perforation of purulent foci: empyema, abscess of the lungs or of the neighboring organs; also in bronchorrhœa.

3. *Muco-purulent sputum* is found most frequently, and has no characteristic points for differential diagnosis. Is seen in severe bronchitis as well as in phthisis pulmonum. In bronchitis pus and mucus are often intimately mixed; in phthisis it frequently consists of single balls which have a gnawed appearance, are surrounded by mucus, and sink to the bottom. This kind of sputum (nummulous, *globosum et fundum petens*) is mostly characteristic of tubercular cavities, but occasionally appears in bronchorrhœa.

Profuse muco-purulent sputum often forms *three layers* on standing: the lowest, pus, then serous fluid, on top foamy mucus. Found most often in bronchiectasis and cavities, but is not pathognomonic.

4. *Serous sputum*, thin, fluid, usually stained a light red, is pathognomonic for œdema of the lungs; its appearance is usually a bad prognostic sign, usually a sign of early death (stertor).

5. *Pure bloody sputum* (hæmoptysis) is expectorated when through some ulcerative process a pulmonary blood-vessel becomes eroded, or when in the pulmonary circulation or in isolated areas (embolism), stasis of high degree is present.

Differential diagnosis as to hæmatemesis (p. 79). Hæmoptysis appears: —

a. Principally in *tubercular phthisis;* sometimes in the first stage (initial hæmoptysis) or in any stage of the subsequent course of the disease; the amount of the coughed-up blood varies from 1 to 2 teaspoons to ½ litre or more. The proof of the presence of other signs of tuberculosis is required to make a positive diagnosis.

b. Less often in abscess of the lung or gangrene.

c. In extensive stasis in the pulmonary circulation, especially in mitral lesions.

d. In hæmorrhagic infarcts of the lung; the source of the embolus must be proven (venous thrombosis, thrombosis in the right ventricle); when possible, the area of dulness of the infarct, or râles, must be elicited.

e. In aneurysm of the aorta which may lead to profuse, often fatal, pulmonary hæmorrhage.

f. In very rare instances, when a dilated vein (varix) of one of the large bronchi bursts in an otherwise healthy person. This diagnosis is justified only when the other causes are excluded.

An hæmoptysis which appears at the time of suppressed menses in a young girl may be regarded as beneficial (vicarious hæmoptysis). One must not neglect the examination for tuberculosis, however.

6. *Bloody-mucous sputum* (like raspberry jelly) is occasionally seen in carcinoma of the lung; when it is more bloody than mucous, sometimes stained yellowish-brown, in the first stage of pneumonia and in hæmorrhagic infarcts.

Saliva stained with blood is often expectorated by hysterical patients, and may lead to error.

Fibrin in quite large quantities is seen in fibrinous bronchitis and pneumonia; if one shakes up the sputum with water in a test-tube, branches of fibrin like those of a tree appear (*bronchial casts*).

In order to decide whether sputum contains fibrin, *Ehrlich's* three-color mixture may be employed. This tri-acid mixture contains acid fuchsin, methyl-green, orange-G. The acid fuchsin is absorbed by the pure albumin present, while nuclein (the albumin of the nucleus) and mucin are stained with the basic methyl-green. If a specimen of sputum is shaken up with this mixture, pneumonic expectoration will be stained red because of the fibrin it contains, the sputum of bronchitis will be a bluish-green because of the mucus and leucocytes it holds. This test may be employed microscopically on dry specimens.

The **smell** and **color** of the sputum are next to be observed. Most sputa have an insipid, sweetish smell. A mouldy odor may accompany material arising from the mouth, the teeth, the nose, and the pharynx. A *sickly, foul odor* is a sign of purulent bronchitis or gangrene of the lung or ruptured putrid abscesses (see below).

A foul odor may come from the œsophagus or nose, which must always be carefully examined as possible sources of putrid sputum.

The **color** is usually a yellowish-green. Sputa of other colors have important bearings: *red color* (see *bloody sputa*).

Rubiginous (rusty) *sputum*, pathognomonic for pneumonia.

Ochre-yellow sputum, pathognomonic for the perforation of foci of liver disease (echinococci, abscesses from biliary calculi, necrotic liver tissue). The color comes from the large quantity of bilirubin.

A similar color is caused by the activity of bacteria: *sputum* of the color of the *yolk of an egg*. The color becomes more intense when exposed to the air, and if a small part of the sputum is transferred to clear sputum, the latter assumes the yellow color in a short time.

Grass-green sputum, characteristic of a slow resolution of a pneumonia, usually indicates a transition to tuberculosis; also seen in pneumonia accompanied by icterus.

The green color may also be caused by bacterial life. This form may be recognized by its imparting its color to clear sputum.

The differently colored sputa which acquire their color from external sources are without diagnostic significance: *blue sputa* in workmen in chemical factories, *red* and *yellow* in workers at the forge, *black* in workers in coal.

Lastly, the **amount** of sputum in 24 hours must be noted. For diagnostic purposes it may be important when there is a very great quantity of secretion (purulent bronchitis, bronchiectasis, tuberculous cavities, perforating empyema, etc.). In many cases the intensity of the pathological process may be estimated by the amount of sputum.

MICROSCOPIC EXAMINATION

A microscopic examination of the sputum must be made in all cases in which the examination of the thorax and the inspection of the sputum have not led to a positive diagnosis (for example, when there is a suspicion of tuberculosis, in the presence of blood-stained, mucous sputum and ill-smelling sputum, etc.).

Important elements are: elastic fibres, fragments of lung tissue, tubercle bacilli, heart-disease cells,[1] eosinophile cells, Leyden's asthma crystals, and Curschmann's spirals. These are some elements which help to secure the diagnosis made: crystals of the fatty acids, hæmatoidin crystals, bronchial casts, etc.

Elements which lack particular diagnostic importance are (Fig. 27):

White blood-cells, present in large quantity in every sputum, often degenerated, often fatty.

Pavement epithelium comes from the mouth and the true vocal cords.

Cylindrical epithelium is not often found; it comes from the nose, the upper pharynx, the lower part of the larynx, and the bronchi.

[1] We have translated the word "Herzfehlerzellen" as above. — THE TRANSLATORS.

The *heart-disease-cells* are large, ovoid or round, with a nucleus which resembles a bubble, and are usually filled with particles of coal, as well as fat and myelin (anthracosis). Although the presence of aveolar epithelium is not characteristic of tuberculosis, the presence of a large quantity of anthracotic cells must awaken a suspicion of the disease.

Isolated red blood-cells are without significance. They are present in abundance in bloody sputum.

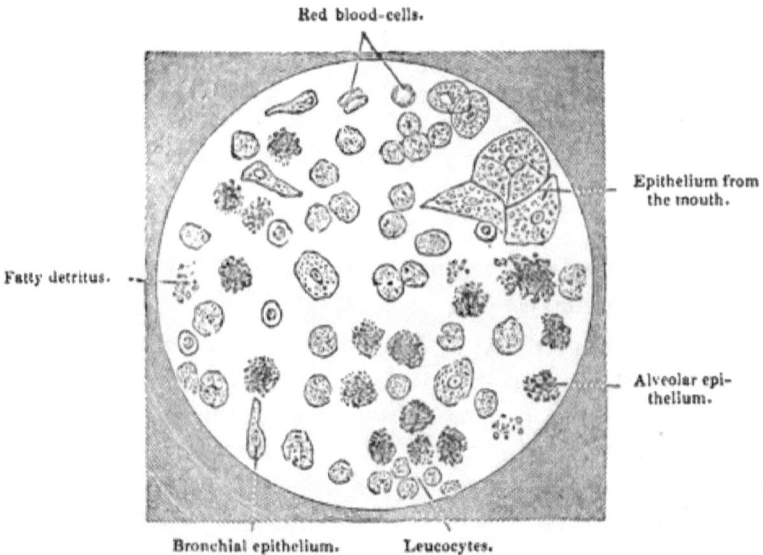

Fig. 27. Morning Sputum in Chronic Bronchitis, containing No Pathological Elements.

Sarcina pulmonum is seen now and then. It has no diagnostic importance.

The ordinary *bacteria* are present in profusion in every sputum, particularly in old sputum, and have no diagnostic value. *Leptothrix* (stained blue with Lugol's solution) are present in gangrene; they also appear in the follicles of the tonsils in the healthy. Lugol's solution contains iodine, 1; iodide of potassium, 2; distilled water, 100 parts.

CHARACTERISTIC ELEMENTS

Elastic fibres, in all destructive processes: tuberculosis, abscess, gangrene. In gangrene not so many fibres are found, for there is a ferment in the sputum which dissolves them. Abscess is rare and is to be diagnosticated from other symptoms (see below), so that, as a rule, the presence of elastic fibres speaks in favor of tuberculosis.

To find elastic fibres, a cheesy particle may be selected from the sputum with a curved forceps. The sputum should be poured on a black plate to render the selection easier. Low powers of the microscope will answer for the preliminary examination. The sputum may be previously boiled in a test-tube with an equal quantity of a 10 per cent. potassium hydrate solution and allowed to precipitate. The sediment is then examined microscopically.

Fragments of lung tissue appear as small black particles, visible to the naked eye, in abscess and gangrene. In foul-smelling sputum their presence speaks in favor of gangrene rather than fetid bronchitis.

Heart-disease-cells are found in the sputum in all diseases in which there is chronic pulmonary congestion, as in mitral disease and in all cases of dilatation of the left ventricle. These conditions frequently superinduce cardiac asthma and secondary emphysema. The heart-disease-cells are desquamated alveolar epithelium containing granular, brownish-yellow pigment.

The source of this pigment is undoubtedly the blood, and may be proven by the hæmosiderin reaction: to a fresh specimen diluted hydrochloric acid and the yellow potassium ferrocyanide (5 per cent. solution) are added. The pigment granules take on a blue color from the iron they contain.

Eosinophile cells appear in the sputum usually in combination with Leyden's asthma crystals. They are large cells with very fine, colorless granulations which stain a beautiful red with eosin. Their presence in abundance seems to bear some relation to attacks of asthma; but nothing definite can be said as to their significance.

The staining of these cells in dry specimens may be best accomplished by the following solution (Dr. Bein): concentrated aqueous solution of methylene blue 50; absolute alcohol and distilled water, of each 24. To this solution is added a piece of eosin the size of a bean. In this solution leucocytes and bacteria are stained blue; the eosinophile granulations, red.

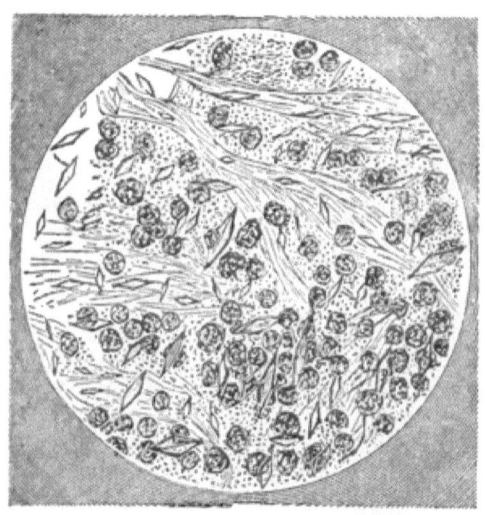

FIG. 28.—ASTHMA CRYSTALS.

Leyden's asthma crystals (Fig. 28), strongly refractive, octahedral figures, are found almost solely in bronchial asthma. Macroscopically, they appear in yellow, sausage-shaped particles. Chemically, they consist of diethylendiamin.

Curschmann's spirals (Fig. 29), also found in the sputum of asthma but seen much less frequently, may be recognized with the naked eye, better with a magnifying glass, as fine thread-like bodies, often contained in sago-like particles of mucus. They are conglomerations of mucus wound in the manner of a corkscrew, with a light central line.

Fibrin may often be appreciated macroscopically. Shaken with water, it may be recognized by its forming bronchial casts shaped like the branches of a tree. Microscopically, it is known by its brightness, its fine streaking, and its homogeneity (staining reaction, p. 144). Fibrin is found in asthma, fibrinous bronchitis, and pneumonia.

Crystals of the fatty acids usually present themselves as curved, colorless needles, frequently in bunches; are easily distinguishable from tyrosin and other crystals by melting when warmed. Are present in gangrene and fetid bronchitis, usually contained in yellow, ill-smelling plugs the size of a pinhead (*Dittrich's* plugs).

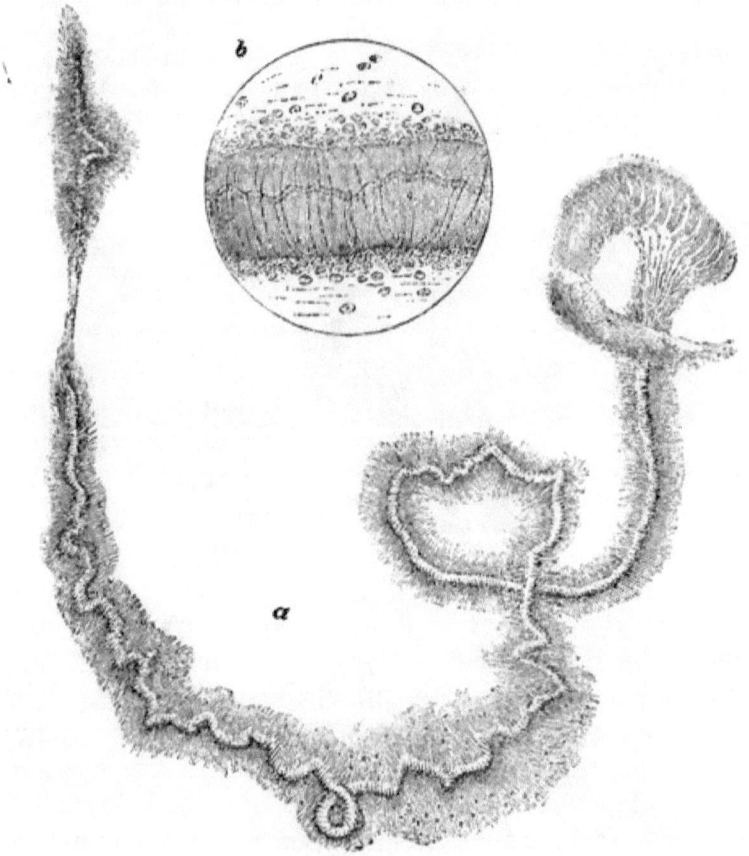

Fig. 29. — Curschmann's Spirals.
a: magnified 80 times. *b*: a piece of *a*, magnified 300 times.

Hæmatoidin crystals are found in old hæmorrhages, especially in abscess and perforating lesions of the liver (ochre-yellow sputum). They occur in brownish-yellow groups, as isolated needles, rhombi, and flakes.

Cholesterin crystals are six-angled, notched plates seldom found in old, purulent sputum (abscess and cavities).

Tyrosin crystals are bunches of needles, in old pus, particularly in the drying pus of perforating empyema.

Small cysts of echinococcus or its hooks are rarely present in the sputum, and when they are, denote echinococcus of the lung or perforation from some neighboring organ (see Chap XII.).

The *cocci of pneumonia* (see Chap. XII.) are present in profusion in every case of genuine fibrinous pneumonia. A simple glance does not suffice, however, to base a diagnosis on, since it is difficult to distinguish them from harmless fungi. They may be recognized with certainty if one emulsifies a quantity of sputum with sterilized water and injects this subcutaneously into a rabbit. If the sputum contain virulent pneumococci, the rabbit will die of septicæmia in from 1 to 3 days, and its blood and the fluid from its enlarged spleen will contain innumerable diplococci. The expectorated material employed must be free from saliva and bronchial mucus and must be positively from the *lungs;* for pneumococci are present in the *mouths* of *healthy persons* and are therefore also called the cocci of sputum septicæmia.

To cleanse the pulmonary sputum, as above mentioned, it is received in a sterilized Petri's dish; it is washed with sterilized water, transferred to another similar dish with sterilized forceps, is again washed in sterile water, etc. After this process is repeated five or six times, it may be safely assumed that the sputum is free from contamination, and may be employed for purposes of culture or inoculation (*Koch's* washing method).

Influenza bacilli (Chap. XII.) are found abundantly in the bronchitis and pneumonia of influenza. The microscopic examination does not always answer, and it may be necessary to make a culture.

Tubercle bacilli. Their demonstration is the keystone of the diagnosis of tubercular phthisis of the lungs. In doubtful cases, especially in the early stages, the proof of their presence is of the highest importance. The finding of tubercle bacilli in the sputum proves with certainty the diagnosis of tuberculosis. A negative result, on the other hand, does not speak absolutely against the diagnosis. For the technic of staining, see Chap. XII. If difficulties present themselves, the method of *Biedert* may offer some advantages. A teaspoonful of sputum, with 3 teaspoonfuls of water and 15 drops of potassium hydrate, are boiled for 2 hours over a sand-bath. The greater part of the pus is dissolved. In the sediment remaining, even a few bacilli may be stained easily and with certainty.

Anthrax bacilli and the club-like ends of *actinomyces* have been occasionally found in the sputum as proof of the existing disease. So have *chains of aspergillus* and the *fungus* of *thrush* in the sputum of pneumonomycosis.

Symptoms of the Diseases of the Lungs

Bronchitis. — Cough and expectoration, frequently pain in the chest, usually no marked emaciation. Physical examination: *no* dulness, vesicular breathing often accompanied by prolonged sharpened expiration, diffuse dry râles. Sputum, in acute cases, glassy, mucous, clinging tenaciously to the bottom of the containing vessel; in chronic cases, muco-purulent, with no characteristic elements. Signs of bronchitis in *one* apex only justify the suspicion of beginning tuberculosis (*catarrhus unius lateris non est catarrhus*). It is worthy of note that signs of acute bronchitis arise secondarily: in cardiac disease (bronchitis of stasis), and in the acute infectious diseases, as in typhoid fever. Prognosis of simple acute bronchitis, with proper treatment, good. Chronic bronchitis leads to emphysema and dilatation of the right heart, and runs its course, frequently, with attacks of bronchial asthma.

Chronic bronchitis may lead, through retention of the secretions, to sac-like dilatations of the smaller air-passages (*bronchiectasis*). Small bronchiectases usually escape diagnosis. In large bronchiectases, there may be a marked stagnation of pus, which undergoes putrefaction (see fetid bronchitis). The sputum, in such cases, is expectorated in the morning frequently by mouthfuls, and arranges itself in three layers. Large bronchiectases give the physical signs of cavities; they usually occupy the lower lobes, and may be concealed by the corresponding compression of the tissues of the lungs.

The etiological irritation which is the cause of bronchitis (dust, etc.) may also evoke interstitial inflammation in the lung, which may lead to formation of connective tissue, and to contraction (interstitial fibrous pneumonia, cirrhosis of the lung, pneumoconiosis). Bronchitis may also, in this way, be responsible for areas of dulness. It must not be forgotten that tubercle bacilli usually find their way into lungs saturated with dust and irritated by it; and in practice it has this significance, — that bronchitis with dulness denotes tuberculosis. The final decision will rest upon the finding of tubercle bacilli.

Emphysema (*volumen pulmonum auctum*). — Shortness of breath and cyanosis, usually cough and expectoration. Barrel-shaped chest. Often dilated veins on both sides of the chest. Lower border of the lung lower than usual; absence or diminution of cardiac dulness. Vesicular breathing diminished in intensity. Frequently, dry râles.

The prognosis depends upon the degree of preservation of the respiratory motility of the lungs; in very advanced cases, the limit of the expiratory position of the lower border of the lungs is from $\frac{1}{2}$ to 1 cm. from their most extended inspiratory position. Some observers reserve the name *emphysema* for those serious cases prognostically, in which the alveoli undergo necrosis, because of the pressure prevailing in them, thus producing a typical (Laënnec's) emphysema, in which air, to a certain extent, penetrates the tissues. If one employs the word *emphysema* in this limited sense, the phrase *volumen pulmonum auctum* would serve well for all those cases of moderate and milder forms which represent clinically a completer conservation of respiratory normality.

Pneumonia (*genuine, croupous, lobar*). — Sudden beginning with chill, pain in the side of the chest, cough. High continuous fever. Rubiginous sputum. Physical signs, — 1st stage (engorgement): tympanitic, slightly dull note over the infiltrated (lower) lobe; crepitant râles; in the 2d stage (hepatization), complete dulness; bronchial breathing. Increased vocal fremitus. 3d stage (resorption or resolution): gradual clearing up of the dulness, the bronchial breathing gradually becomes vesicular; fine and coarse râles may be heard. Prognosis: usually favorable; crisis between the 3d and 11th days, resolution of the exudate (disappearance of dulness) in 1 to 4 weeks. In weakened persons, the resolution may require months (as much as $1\frac{1}{2}$ years) (delayed resolution). Fever of remittent type lasting longer than 11 days points to complication or sequelæ, especially pleurisy (empyema); rare issues are carnification (formation of granulation tissue in the exudate with subsequent connective tissue formation and contraction), cheesy degeneration, abscess, gangrene. Prognosis for healthy young persons usually good; in individual cases the prognosis depends particularly upon the condition of the heart (pulse) and the involvement of the mental faculties. Prognosis grave in drunkards, aged people, cardiac cases, and persons with kypho-scoliosis.

Broncho-pneumonia (*secondary, catarrhal, lobular*). — Follows bronchitis, usually in acute infectious diseases: moderate dulness often with a tympanitic sound, with bronchial breathing and moist râles, over circumscribed areas most frequently involving both lower lobes. Remittent fever of long continuance, sputum usually muco-purulent, colorless. Appears chiefly in children and the aged, frequently in stuporous patients who aspirate particles of food (aspiration pneumonia).

The prognosis of a broncho-pneumonia is always grave; recovery is possible, though most frequently death occurs as a result of exhaustion, tuberculosis, or gangrene.

Dry pleurisy (*pleuritis sicca*) is diagnosticated by the presence of localized *friction sounds*, which are equally loud on inspiration and expiration, and may frequently be felt. Pain and usually fever. Not rarely caused by exposure to cold. Often present in tuberculosis.

Pleurisy with effusion (*pleuritis exsudativa*). — Begins suddenly with a chill and pain in the side, or the pain may come on gradually; frequently low degree of dyspnœa. Fever irregular, remittent. Absolute dulness (flatness) below posteriorly, respiratory murmur and pectoral fremitus diminished or absent. Above the border of dulness, on account of the compression of the lung, often a tympanitic, slightly dulled percussion note, bronchial breathing, very fine râles (atelectasis). Apex-beat and cardiac dulness frequently displaced. When the effusion is on the left side, the semilunar space is usually diminished.

The upper border of the dulness is situated differently, according as the patient lay or walked about during the formation of the effusion. In the first instance, especially among the better classes, the border of dulness runs diagonally from above posteriorly to below anteriorly; if the patient had been walking about (as is frequently the case among hospital patients), the upper border is almost a horizontal line. During the stage of resolution the upper border often forms a convex curve pointing upward, the highest point of which lies in the side of the chest (the curve of *Ellis* and *Damoiseau*). By the change of position of the patient the border of dulness fixed by inflammatory adhesions is altered slowly or not at all.

After a pleuritic effusion is diagnosticated, the nature of the exudate must be determined, whether it is *serous* (simple pleurisy), *purulent* (empyema), or *hæmorrhagic* (usually dependent upon a malignant new growth). Although the patient's condition of strength, the character of the fever, the pulse, and the respiration must be considered in making a differential diagnosis, an (aseptic) *exploratory puncture* with a hypodermic syringe must be undertaken to settle the diagnosis. It should never be omitted in any case of pleural effusion.

In considering *serous* effusions, it must be noted that it is frequently evoked by a primary disease of the lungs (particularly tuberculosis, next in importance pneumonia, infarct, gangrene, abscess). One should not be satisfied, therefore, with the diagnosis "serous pleurisy," but should examine the lungs, and, if necessary, the sputum, in order to ascertain the etiological factors of the pleurisy. It is important to remember in this connection that many signs of tuberculosis of the lungs lose their importance when found above a pleural effusion; for dulness, bronchial breathing, and crepitant râles may arise as well from compression of the lung.

Aside from diseases of the lung, other things that may produce pleurisy are: diseases of the heart and kidneys, inflammatory conditions of the liver, circumscribed peritonitis, particularly appendicitis, possibly lues; as complications of acute infectious diseases, pleuritis not infrequently arises, just as do inflammations of other serous surfaces (pericarditis, endocarditis).

In *purulent* effusions, the treatment and prognosis are mainly dependent upon the determination of the etiology; aside from the history, many cases require a bacteriological examination of the bacteria contained in the pus obtained from the exploratory puncture.

Pneumococci in the pus denote *meta-pneumonia*, tubercle bacilli *tubercular* empyema. Streptococci and staphylococci give no certain evidence as to the origin of the disease; in tubercular disease of the lungs, empyema dependent upon streptococci and staphylococci may arise. Bacilli of putrefaction are found in empyema in gangrene of the lungs or in embolic infarcts of rotten purulent processes, especially in puerperal infections. *The continued absence of bacteria in the pus speaks for tuberculosis.*

In many cases during and after the resolution of pleural effusions, adhesions between the two layers of the pleura spring up with the formation of thick connective tissue (*pleuritis retrahens*). The *pleuritic thickenings* manifest about the same physical signs as the effusion (dulness, diminution of the respiratory murmur, and the vocal fremitus), yet they differ in this respect, that they show signs of retraction and of tension on neighboring organs (flattening and drawing in of one side of the thorax, displacement of the heart, or increased size of the semilunar space).

Phthisis pulmonum. — The first stage may give evidence of no physical changes in the thorax. The suspicion of tuberculosis is aroused by various uncertain symptoms: a hacking cough, little expectoration, headache, easy fatigue, loss of appetite, gastric disturbances, loss of flesh and strength; an hereditary taint and the habitus paralyticus make the aspect of the case more serious. The (early) diagnosis can be made with certainty only by the presence of tubercle bacilli in the sputum.

A rise of temperature following an injection of *Koch's* old tuberculin may be counted as one of the signs which enable an early diagnosis to be made. Although convalescents and weakened persons with healthy lungs may also show a rise of temperature after small doses, still the fever induced by the subcutaneous injection of from 1 to 5 mg. of tuberculin speaks very much in favor of tuberculosis. These injections are not without some danger, however, and are scarcely to be commended for use in private practice.

It is important, for diagnostic purposes, to discover possible sources of infection: residence with tubercular patients (wife or husband, brother or sister, other occupants of the same house).

The first appreciable physical signs are: distinct dulness over *one* apex; vesicular respiration with sharpened, prolonged expiration or broncho-vesicular breathing; fine, non-metallic or bronchitic râles.

In a further advanced stage, marked emaciation, profuse cough and expectoration. Sputum large quantity, muco-purulent, often in little balls; contains elastic fibres and tubercle bacilli. Intense dulness over the apex and below the clavicle, bronchial breathing, profuse, coarse râles with something of a metallic ring.

In the final stage, extreme emaciation, very profuse cough and expectoration, sputum in balls and falling to the bottom of a vessel. Physical examination: in part extensive areas of dulness, in part a loud tympanitic note (even the lower lobe is sometimes attacked), bronchial breathing, tinkling, coarse râles, in places change of sound in the percussion note.

Prognosis in the very beginning with the possibility of thorough treatment, inclined to be good; in the more advanced stages, usually bad. Complication, in the beginning, usually pleurisy; other possible complications: pneumothorax, miliary tuberculosis, tuberculosis of other organs (larynx, intestine, tubercular meningitis, peritonitis, etc.), general amyloid degeneration.

Pneumothorax.—Rarely seen in the healthy (trauma, fracture of ribs, over-exertion); usually secondary to phthisis, gangrene, abscess, perforating empyema, emphysema. The physical signs are:—

Dilatation of the affected side and its immobility in the existing dyspnœic breathing. Percussion: abnormally loud, deep note (usually not tympanitic), metallic note when the hammer and pleximeter are used. Auscultation: if the cavity-opening is closed, no respiratory murmur; if it is open, amphoric breathing. A fluid effusion usually takes place soon; sero- or pyo-pneumothorax (exploratory puncture); above the fluid, dulness with no respiratory murmur and no vocal fremitus; instantaneous change in the borders of the dulness on change of position. Metallic splashing sound on shaking the thorax from side to side, audible at quite a distance (*succussio Hippocratis*). The prognosis depends upon the primary disease and the possibility of operative interference. Prognosis in the healthy, good.

Fetid bronchitis is diagnosticated when the sputum is foul-smelling (nose and throat not being involved), when no characteristic elements aside from the fetid plugs are formed in the expectoration, and when there is *no dulness* over the lungs and the signs of bronchitis only are present.

The prognosis depends upon the intensity of the bronchitis or upon the presence of bronchiectasis and upon the general symptoms evoked by the putrescence. Fetid bronchitis, with no septic manifestations, usually offers a good prognosis. Fetid bronchitis frequently leads to large bronchiectases which may firmly compress the surrounding tissue. In this manner, a putrid bronchitis may call forth dulness (in the lower lobes).

Gangrene of the lungs is diagnosticated from the putrid expectoration, which contains fragments of lung tissue in addition to fetid plugs (p. 149), and the physical demonstration of the necrotic area of lung: dulness, bronchial breathing, moist râles.

The diagnosis must also establish the *cause* of the gangrene, which modifies the prognosis: trauma (usually great pressure), pneumonia (frequently after influenza), embolus, perforating putrid abscess of the bronchial glands, perforating putrid empyema, extension of gangrenous foci from the œsophagus, vertebræ, intestines, liver.

As far as the prognosis is concerned, there must be determined: (1) the extent of the local disease: in circumscribed gangrene without perceptible formation of cavities, prognosis inclined to be good; in diffuse gangrene with formation of cavities, prognosis bad; (2) the cause of the gangrene: injury and pneumonia, empyema, bronchial abscesses, give comparatively a good prognosis; suppurating emboli, extension from the œsophagus, vertebræ, etc., usually a bad prognosis; (3) the general symptoms: signs of severe infection (very rapid pulse, delirium and collapse) are of evil omen.

Hæmorrhagic infarct of the lung is diagnosticated when in conditions which accompany the possibility of local *thrombosis* (puerperium, marasmus, wounds, decubitus, etc., especially *dilatation of the right heart*) there are sudden stitches in the side, cough, and bloody sputum often accompanied by fever. The diagnosis is made more certain by the demonstration of a circumscribed infiltration (dulness, diminished or bronchial breathing, râles), frequently an additional pleural effusion. The prognosis depends upon the cause of the embolus and the bodily strength; small infarcts, not infected, are readily absorbed.

Abscess of the lung is diagnosticated from purulent sputum containing elastic fibres *without* tubercle bacilli, with remittent fever, when the cause of the abscess (pneumonia, infected infarct, injury) may be simultaneously proven and the signs of infiltration or of a cavity are present. Prognosis dependent upon the cause and the general manifestations; a favorable outcome lies in perforation into a bronchus and subsequent healing.

Tumor of the lung (carcinoma or sarcoma) produces local dulness, over which may usually be heard bronchial breathing or râles. The cutaneous veins in the region of the dulness are usually dilated and frequently the veins of the corresponding arm are involved. In many cases, a sputum like raspberry jelly is expectorated. Sometimes a hæmorrhagic effusion into the pleura occurs, often a swelling of the axillary glands. Cachexia sets in during the course of the disease.

Echinococcus of the lung can be diagnosticated only when the signs of a tumor are present and when echinococcus cysts are found in the sputum.

Syphilis of the lung must be considered from the diagnostic point when a diffuse infiltration or contraction presents itself, when tertiary syphilis is known to be present and tuberculosis is excluded. The proof of the diagnosis lies in the successful issue of specific treatment.

Actinomycosis of the lungs. — Signs of infiltration and cavity formation in the lungs, with considerable secondary pleurisy. The diagnosis is made more positive by the demonstration of the presence of the ray fungus in yellow granulations contained in the sputum.

CHAPTER VII

DIAGNOSIS OF THE DISEASES OF THE CIRCULATORY SYSTEM

For the anamnesis, the following things must be considered: 1. The previous life of the patient: *excessive physical exertion* and great *psychical excitement* cause idiopathic cardiac disease. Too luxurious eating and drinking produce increased blood pressure followed by arterial sclerosis and heart-disease. Too good living produces obesity (*fatty heart*). *Alcoholism* evokes cardiac weakness; too much smoking, neurotic conditions of the heart. 2. Previous diseases: *acute articular rheumatism*, not quite so frequently all the other acute infectious diseases (scarlatina, erysipelas, malaria, etc.), leads to endocarditis or myocarditis. *Syphilis* may be the cause of a myocarditis. 3. Previous symptoms of heart and kidney diseases possibly present.

The diagnosis of diseases of the heart is supported by subjective symptoms: abnormal sensations in the cardiac region, palpitation of the heart, a sense of anxiety; by the presence of dyspnœa, cyanosis, and œdema, and by the results of the physical examination of the heart and blood-vessels.

Complaints of cardiac difficulties (palpitation, sense of anxiety, etc.) without real dyspnœa, without cyanosis and œdema and in the absence of physical anomalies, are to be referred to *nervous* affections of the heart.

Tachycardia, increased rapidity of the pulse, is often seen in nervous affections of the heart in consequence of excitement, excesses, etc., and as a result of certain digestive disturbances and sometimes without appreciable reason (cf. p. 171).

Angina pectoris is a severe pain in the region of the heart coming on in attacks, usually with pain extending toward the left arm and accompanied by the most exquisite sensations of anguish and anxiety. It may appear in any severe cardiac disease, especially in sclerosis of the coronary arteries (*angina pectoris vera seu Heberdenii*). Painful attacks like those of angina pectoris may arise in neurasthenics (*angina pectoris neurasthenica*); in these cases there are usually other symptoms on the part of the vaso-motor system present, such as alternate flushing and pallor of the face and arms. In every case, angina pectoris is to be regarded as a serious matter which demands the closest examination of the circulatory apparatus.

For a discussion of dyspnœa, cyanosis, and œdema. see pp. 10, 125.

Cardiac asthma is the name given to attacks of dyspnœa in cardiac disease, which last for hours, more rarely for days, and which are followed by intervals of freedom; it may appear whenever there is dilatation of the left ventricle. The differential diagnosis as to bronchial asthma rests upon the *establishment of the dilated ventricle* and the small, rapid, irregular pulse.

In attacks of asthma of doubtful etiology the greatest value must be placed on the examination of the heart, particularly the palpation of the apex-beat and the feeling of the pulse. In *bronchial asthma the heart is healthy*. In other respects, the two diseases may simulate each other closely, since in all weakened conditions of the left heart, a bronchitis due to *congestion* and a secondary *emphysema* may arise (expansion of the lung due to distention of the pulmonary vessels). A characteristic difference is often seen in the character of the sputum (see p. 147). In bronchial asthma the expectoration is tenacious, glassy, mucous, with whitish-yellow lumps, containing, on microscopic examination, crystals, spirals, and eosinophile cells; the sputum of cardiac asthma usually shows the presence of blood and approximates in appearance that of œdema of the lungs. Microscopically, alveolar epithelium is often to be seen.

The *œdema* of cardiac disease begins at the ankles and slowly ascends; it attacks last the hands, arms, and face; the œdema of Bright's disease usually begins in the face.

Albuminuria in heart-disease appears only when a high degree of stasis is present; at the same time the urine is scanty and of high specific gravity (see below).

The objective examination of the heart consists of inspection and palpation, percussion and auscultation of the heart and great vessels, examination of the radial pulse and of the urine.

INSPECTION AND PALPATION

The *position* and *force of the impulse of the heart* and of the *apex-beat* must be determined.

The *impulse of the heart* denotes the impulse of the systole, the elevation of the entire cardiac area; the *apex-beat* is the visible and palpable expansion of the intercostal space furthest to the left, external and inferior.

1. *Position of the apex-beat.* In healthy persons, the hand placed over the cardiac area feels a weak, systolic impulse; the apex-beat is to be felt in the 5th intercostal space midway between the parasternal and mamillary lines.

In children the apex-beat is frequently somewhat higher, in the aged one intercostal space lower; in deep inspiration the apex-beat sinks a little lower. With the patient lying on the left side the apex-beat may be moved a finger's breadth to the left; this is especially true of rapidly emaciating persons, when the patient turns to the right side the apex-beat soon recovers its normal position.

Permanent displacement of the apex-beat is of great diagnostic importance; it denotes either *dilatation* or *compression* of the heart.

Displacement of the apex-beat toward the left signifies (a) *dilatation of the left side of the heart*, (b) *dislodgment* of the entire heart toward the left; in the latter cases a pleural effusion or pneumothorax exists on the right side, in rarer instances a tumor may be demonstrated, or there may be a retracted pleura on the left side.

Displacement of the apex-beat toward the right always depends upon a *dislodgment* of the heart, through a pleuritic effusion or pneumothorax on the left side, or a retracted pleura on the right side.

Displacement of the apex-beat downward may occur through hypertrophy of the left ventricle, less often aneurysm of the aorta, or unusually deep attachments of the diaphragm.

Displacement of the apex-beat upward takes place only when the diaphragm is pushed up in consequence of extraordinary distention of the abdomen (ascites, meteorism, dilatation of the stomach, tumors, pregnancy).

2. *Force of the cardiac impulse and apex-beat.* A diminution in the force of the cardiac impulse and apex-beat to a point at which neither can be felt may occur: (1) in very fat persons; (2) when the lung overlaps the heart: emphysema; (3) when the pericardium is distended with fluid, more rarely when there is a tumor of the pericardium; (4) in all *weakened conditions of the heart.*

In many healthy persons, however, the apex-beat is never felt because it strikes against the rib instead of in the intercostal space.

Increased force of the cardiac impulse and apex-beat. An "elevating" impulse is imparted to the hand. It is present: (1) in *physiologically increased cardiac activity,* when there is psychical excitement, great exertion, in fever; (2) in *hypertrophy of the heart;* (3) often in *dilatation of the heart* when the apex-beat is also displaced outward.

While the diminished force of the apex-beat frequently speaks for cardiac weakness in persons who are not very fat and not emphysematous, its increased force by no means indicates an increased cardiac strength. *Martius* established the fact that the appreciation of the cardiac impulse depends not only upon the functional activity of the heart, but upon the cardiac area exposed to the thoracic wall. According to Martius, the systole of the ventricle may be divided into two periods. It first contracts with closed aortic valves; by this manœuvre its form is changed in a typical way, producing the cardiac impulse, but its volume is unchanged (*period of closure*). During the second period of the systole the aortic valves open, the volume of the ventricle is diminished (*period of expulsion*). This is the explanation, too, of the fact that very weak, dilated hearts with a small pulse often give a decided impulse; during the period of closure a much larger cardiac volume than normally is impelled against the chest-wall, and much less blood is sent into the aorta during the period of expulsion than from a healthy heart.

Sounds or *murmurs* felt by palpation have the significance as when appreciated by auscultation. The *thrill* palpable over *stenosed* cardiac valves is particularly worthy of note. At the apex a presystolic thrill is characteristic of mitral stenosis; at the right edge of the sternum in the 2d intercostal space a systolic thrill is characteristic of aortic stenosis. (Aneurysm must not be forgotten, however.)

A bulging of the cardiac area speaks for dilatation and hypertrophy of the heart or pericardial effusion, although rhachitic changes in the bones must be borne in mind.

A systolic retraction of the 5th intercostal space by the apex-beat occurs only when the two layers of the pericardial sac are adherent in consequence of chronic pericarditis; the pulsus paradoxus is often present (p. 173).

Visible pulsations (synchronous with the radial pulse): over the aorta or pulmonary artery denote aneurysm or an infiltration of the corresponding lobe of the lung; in the *epigastrium*, frequently of no diagnostic significance (when the diaphragm lies low), more often depending upon dilatation of the right ventricle. Visible pulsations of the *liver*, synchronous with a venous pulse, have the same meaning as the actual venous pulse (tricuspid insufficiency).

Venous pulsations, visible at the bulb of the jugular vein or in the jugular vein if the valves of the bulb are insufficient, are either synchronous with the systole of the heart (actual venous pulse, presystolic-systolic), or they precede the cardiac systole (diastolic-presystolic) (cf. p. 176).

An actual venous pulse is the principal sign of tricuspid insufficiency; the presystolic venous pulse is often seen in conditions of venous congestion without valvular insufficiency

Percussion of the Heart

Normal limits of cardiac dulness. The internal border runs along the left edge of the sternum; the external border forms a somewhat convex arch directed outward from the 4th costal cartilage to the 5th intercostal space, between the mamillary and parasternal lines (apex-beat). The upper border lies at the lower edge of the 4th rib, the lower border upon the 6th rib, although this can not always be accurately determined because of the encroaching liver dulness.

The limits described are those of *absolute* flatness, *i.e.*, within these limits the dulness is intense; beyond these limits the so-called "relative dulness" lies, above as high as the upper border of the 3d rib, to the right as far as the median line; but this "relative dulness" is normally not very intense.

In children the area of cardiac dulness is somewhat greater, in the aged somewhat smaller. Every deep inspiration diminishes the area. With the patient lying on the left side, the external border moves outward about one finger's breadth.

The **increased area of cardiac dulness** is one of the principal signs of advanced heart-disease. Lateral extension of the dulness usually denotes dilatation of the ventricle. Dilatation is the *second stage* in most cardiac diseases, and arises from hypertrophy.

1. *Extension of the cardiac dulness to the left* over the mamillary line denotes dilatation of the left ventricle; this is induced by aortic insufficiency or stenosis, by mitral insufficiency, and by the causes of idiopathic heart-disease (see p. 176).

2. *Extension of the cardiac dulness toward the right* beyond the left border of the sternum points to dilatation of the right ventricle, although a similar note evoked over the lower half of the sternum may be due to a mere collection of fat; dilatation of the right ventricle arises in mitral stenosis and insufficiency, in valvular lesions of the right

side of the heart as well as in emphysema, kypho-scoliosis, retracted pleuræ.

3. *Simultaneous expansion of the cardiac dulness toward both sides and upward* denotes a fluid effusion into the pericardium (pericarditis or hydro-pericardium). The area of dulness forms an equilateral triangle the apex of which lies in the 3d to the 1st intercostal spaces.

In every case of enlargement of the area of cardiac dulness, it must be decided whether there is an *actual* dilatation (increase of volume), or whether there is (1) a displacement of the entire heart, (2) a withdrawal of the lung covering the heart so that a greater cardiac area lies directly against the chest-wall. The cardiac dulness is changed in place by pneumothorax, pleuritic effusions, tumors, retracting processes of the pleuræ and lungs; the area is freed from a covering of pulmonary tissue in retraction of the lung; the heart is likewise brought nearer the chest-wall by upward pressure of the diaphragm (ascites, pregnancy, etc.). In cases of transposition of the viscera the entire heart lies on the right side, the liver on the left.

Hypertrophy of the heart is usually *not* demonstrable by percussion; only after dilatation is added to the hypertrophy can one make the percutory demonstration.

Hypertrophy of the left ventricle is diagnosticated by the powerful apex-beat, accompanied by an abnormally high tension of the radial artery, accentuation of the systolic mitral and the diastolic aortic sounds.

Hypertrophy of the right ventricle is diagnosticated by the abnormal accentuation of the diastolic pulmonic sound.

Diminution or *absence* of the cardiac area of dulness is found when the over-distended lung covers the heart (emphysema).

The entrance of air into the pericardium (*pneumo-pericardium*) produces a tympanitic or metallic percussion note over the area of cardiac dulness; this is a fatal, very rare phenomenon caused by perforation of an ulcer of the stomach or the contents of a lung cavity into the pericardium.

Dulness over the upper part of the sternum, or directly next to it, is diagnostic of aneurysm of the arch of the aorta or of a mediastinal tumor; in very rare cases, of an enlarged thymus gland or of a sub-sternal goitre.

Auscultation of the Heart

The auscultation of the heart shows whether or not there are valvular lesions present; these lesions are recognized by characteristic murmurs. Clear heart-sounds denote the intactness of the valves; but despite this fact, the heart may be diseased, hypertrophic, or dilated. *Dilatation and hypertrophy of the heart in the presence of clear tones rest upon disease of the cardiac muscle* (idiopathic heart-disease).

Normal and Intensified Sounds

The sounds of the *mitral valve* are auscultated at the apex, those of the *tricuspid* at the right border of the sternum at the 5th and 6th costal cartilage; the sounds of the aortic valves are best heard at the right border of the sternum in the 2d intercostal space; those of the pulmonic valves at the left border of the sternum in the 2d intercostal space.

Over each valve are heard a systolic sound during the contraction of the ventricles and a diastolic sound during the dilatation of the ventricles.

Over the mitral and tricuspid valves only one sound is heard, the systolic, produced by the tension of the valves and the contraction of the muscles of the ventricles; the diastolic sound is carried away by the aortic in the one case, by the pulmonic in the other. Over the arterial openings, two sounds arise, the systolic, produced by the tension of the expanded vessels, the diastolic, through closure of the valves.

Over the mitral and tricuspid valves, the systolic sound is normally somewhat louder than the diastolic; over the aorta and pulmonary artery the diastolic is normally somewhat louder than the systolic.

Abnormal *intensification* of the mitral systolic sound in hypertrophy of the left ventricle and in physiologically increased activity of the heart (exertion, excitement); also in fever.

Abnormal *diminution* of the mitral systolic sound in all weakened conditions of the left ventricle (often dilated).

Abnormal *intensification* of the 2d pulmonic sound denotes hypertrophy of the right ventricle.

Abnormal *intensification* of the 2d aortic sound denotes hypertrophy of the left ventricle.

A *musical timbre* in the heart-sounds does not allow of essential diagnostic discrimination; it is usually produced by an increased tension of the flaps of the valves.

A *metallic sound* of the cardiac sounds (often to be heard at some distance) proves the presence of large spaces of air near the heart; also heard in cavities of the lung, dilatation of the stomach, and in the very rare, fatal cases of the entrance of air into the pericardium (pneumo-pericardium).

Reduplication of the heart-sounds is of little diagnostic value. It occurs in the healthy and is particularly frequent in the systolic sounds heard at the apex in hypertrophy following contraction of the kidney. Reduplication of the diastolic sound in consequence of mitral stenosis.

MURMURS

Systolic and *diastolic* murmurs are distinguished; their nomenclature depends upon the fact whether or not they are synchronous with the cardiac impulse (or pulse) or not. A diastolic murmur which immediately precedes the impulse of the heart is called presystolic. A murmur occurs simultaneously with a heart-sound, or after it, or with no relation to the sounds.

The murmurs are best heard in a direction perpendicular to the current of blood which produces them. Auscultation in mitral insufficiency is therefore practised in the 2d left intercostal space; in aortic insufficiency the diastolic murmur is best auscultated at the middle of the sternum or at the left border of the sternum in the 3d intercostal space.

A *systolic* murmur over the *mitral* denotes insufficiency of the mitral. This murmur may depend upon anatomical changes (endocarditis); it may, however, be functional or *accidental*.

Accidental murmurs are produced by a turning over of the borders of the valve in consequence of undue stretching of the papillary muscles or by relative insufficiency in consequence of dilatation of the ventricle. Accidental murmurs are soft, blowing, usually heard only as *systolic* murmurs.

A systolic murmur at the apex is regarded as functional, if the patient has *fever*, is *anæmic* or *poorly nourished*, and the murmurs disappear with time. It may be referred to an endocarditis, when a sufficient etiology can be gathered (especially articular rheumatism) and other symptoms of valvular lesions are present (accentuation of the 2d pulmonic, dilatation of the right ventricle, etc.).

Diastolic (*presystolic*) murmur over the *mitral* denotes mitral stenosis.

Systolic murmur over the *aorta* is diagnostic of aortic stenosis.

Diastolic murmur over the *aorta* signifies aortic insufficiency caused by endocarditis or arterio-sclerosis.

If *two* murmurs are heard, the greatest importance is to be attached to the diastolic.

Diastolic murmurs are but rarely accidental, while systolic murmurs frequently rest upon a functional basis.

The force and the character of the murmur offer a small basis as to the prognosis of the valvular lesion.

The force of the murmur is only in part dependent upon the severity of the anatomical change; of greater importance are the rapidity of the blood-current, the smoothness or roughness of the walls of the valves. The character of murmurs is described as blowing, grating, scratching, etc.

Pericardial friction sounds are not synchronous with the heart's beat, seem to be nearer to the ear on auscultation than endocardial murmurs, often heard at irregular intermissions (puffing murmurs). They prove the presence of fibrinous deposits on the pericardium (*pericarditis fibrinosa*). They are independent of the respiration, but are influenced by very deep inspiration.

Extra-pericardial friction sounds, arising between the pleura and the external layer of the pericardium, have the character of pleuritic friction sounds, usually of a crackling nature, are dependent for their production upon the respiration, and disappear when the breath is held.

Auscultation of the Vessels

The auscultation of the vessels sometimes helps to establish the diagnosis of a valvular lesion.

The systole of the heart corresponds to the diastole of the vessels; heart's systole = vessel's diastole; heart's diastole = vessel's systole.

The *carotid* is best auscultated at the inner margin of sterno-cleido-mastoid muscle at the level of the thyreoid cartilage; the subclavian in the outer portion of the supra-clavicular fossa.

Over the carotid and the subclavian one hears normally two sounds, the first (cardiac systole) arises from tension of the wall of the vessel; the second (cardiac diastole) is transmitted from the aortic valves.

In *aortic insufficiency* a sawing, systolic murmur is heard over the carotids, because of the extraordinarily sudden tension of the walls of the carotid at the instant of the entrance of the current of blood; the second sound is missing, however, as it is not formed by the aortic valves. A systolic murmur may frequently be heard by transmission from the heart in aortic stenosis, mitral insufficiency, and general arterio-sclerosis.

The *more distant* arteries may also be auscultated (the femoral in the groin, the brachial at the bend of the elbow, the radial above the wrist). In healthy persons no sounds or murmurs are heard in these vessels; by pressing the stethoscope upon the artery a murmur produced by pressure may be elicited (arterial diastolic murmur: by very hard pressure, this is perceived by the ear as a clear sound). An *abnormal sound* is found in aortic insufficiency even in the smaller vessels (in the palm of the hand, in the forearm, etc.). A *double tone* is heard in the *femoral* in aortic insufficiency, mitral stenosis, pregnancy, lead colic.

Actual *murmurs*, audible without pressure, heard over peripheral arteries prove the presence of an aneurysm, and are usually palpable.

Normally, nothing is heard over the veins. The *jugular* vein may be auscultated at the outer margin of the sterno-cleido-mastoid muscle at the level of the thyreoid cartilage.

In all cases of anæmia and chlorosis, a loud humming murmur may be heard over the jugular vein, which sounds loudest when the patient turns his head to the opposite side. Over the femoral vein a murmur may be heard only in anæmia of great intensity.

The Pulse

The **frequency** of the pulse in healthy adults varies from 60 to 80 beats a minute, in children from 100 to 140, in the aged from 70 to 90.

Slowing of the pulse (brachycardia, pulsus rarus) is of diagnostic value only under certain circumstances. It is found in the most widely different conditions, evoked by irritation of the vagus, or paralysis of the sympathetic, or irritation or paralysis of the cardiac centres. Brachycardia in exhaustion, after a crisis or at the beginning of convalescence, is especially noteworthy. It is found in *meningitis* (pressure on the brain), in *jaundice* (action of the biliary acids), and in *colic* in which it offers a point in differential diagnosis between it and peritonitis. A slowing of the pulse occurs most frequently in *stenosis* of the aortic and mitral valves among the cardiac diseases, as well as in some idiopathic heart-diseases (coronary sclerosis) and as a result of the action of digitalis.

Rapidity of the pulse (tachycardia, pulsus frequens) evoked by paralysis of the vagus, irritation of the sympathetic, or affections of the cardiac ganglia. Normal after physical exertion, psychical irritation, and after eating; pathological in all febrile diseases (with each degree of temperature the pulse is increased 8 beats a minute), common in convalescence from them; in all febrile diseases which lead to consumption (phthisis, anæmia, etc.).

Excessive rapidity (over 160) is a sign of great weakness of the heart (collapse).

Tachycardia is a sign of disturbed compensation in cardiac diseases and is often proportional to the intensity of the disturbance. Tachycardia is also one of the principal symptoms of neurotic disease of the heart, and when it appears in attacks, forms a particular disease (paroxysmal tachycardia).

Tachycardia with exophthalmos, goitre, and tremor of the fingers, frequently associated with general cachexia, forms the symptom-complex of *Basedow's* disease.

The **rhythm** of the pulse. Irregularity of the pulse (arythmia) is present in many cardiac diseases without offering points for differential diagnosis. Arythmia of mild character is not uncommon among nervous people, after excitement, excesses, gastric disturbances, in constipation; sometimes no reason can be assigned for it. Although arythmia always demands a careful examination of the heart, the diagnosis of a cardiac disease should not be made from this symptom alone.

Embryocardia is the equalization of the systole and diastole through the absence of the normal pause following the diastole. The heart-beats sound like the tick-tack of a clock, like the fœtal heart-sounds. Embryocardia is often the expression of a very weak heart.

Characteristic types of irregular pulse are: *Pulsus alternans:* two heart-contractions correspond to one pulse-beat, or the second pulse-beat is but feebly to be felt. *Pulsus bigeminus:* every third beat of the pulse is omitted. *Pulsus trigeminus:* every fourth pulsation is omitted. No diagnostic conclusions can be reached from these irregularities.

Pulsus paradoxus: in deep inspiration, the pulse becomes small or disappears; appears in adhesions of the layers of the pericardium, mediastinitis with cicatrisation or adhesions, mediastinal tumors, stenosis of the air-passages.

Unequal rebound of the pulse at symmetrical arteries, or a *delayed* pulse in corresponding arteries, is a symptom of aneurysm. The former condition may be due, however, to congenital anomalies of the arteries.

Velocity of the pulse (quick or slow): the pulse is rapid or slow according to the rapidity or slowness of the expansion and collapse of the arterial wall. A *rapid pulse* (*pulsus celer*) is found in all conditions of increased cardiac activity, particularly hypertrophy of the left ventricle. It is characteristic of *aortic insufficiency* (*pulsus celer et altus*), contracted kidney, *Basedow's* disease, etc. A *slow pulse* (*pulsus tardus*) is seen in the aged, in aortic and mitral stenosis, in aneurysms.

The **tension** of the pulse (large or small, high or low): the height of the pulse-wave depends upon the strength of the heart, the amount of blood in the arteries, and the tension of the arteries. A high-tension pulse is present in fever, hypertrophy of the heart, especially in aortic insufficiency; a small pulse is a sign of cardiac weakness, characteristic among valvular lesions, of stenosis.

Hardness of the pulse (hard or soft, *durus* or *mollis*): this quality depends upon the *tension* of the arterial walls and is proportional to the strength which must be used by the examining finger to compress the pulse. Hard pulse in hypertrophy of the left ventricle (wiry in contracted kidney), as well as in tetanic contraction of the arterial muscles. Soft pulse in anæmia and fever. Hard pulse in *arterio-sclerosis* due to *deposits of lime* in the walls of the arteries: the artery may be rolled under the finger.

SPHYGMOGRAPHIC TRACINGS (Figs. 30-33)

The purpose of sphygmographic tracings of the pulse-curve is to reach precision in the recognition of the changes in the pulse by *objective* means. A diagnosis may be supported by such tracings.

An ascending and a descending limb is seen in sphygmographic curves. Elevations arising from the ascending limb are called *anacrotic*; arising from the descending limb, *katacrotic*. In the normal pulse of adults, the ascending limb rises almost perpendicularly. Anacrotic elevations appear only in disease of the heart or arteries, the expansion of the arteries occurring in intermissions. The descending limb has normally a decided elevation: the *dicrotic wave*, produced by the falling back of the blood against the aortic valves; and several minor waves (waves of elasticity, tidal and *postdicrotic*) evoked by the elasticity of the arterial wall. The tidal waves are marked when the walls of the artery are under tension, as in lead colic. At the same time the dicrotic wave is small. The tidal wave is very small or disappears altogether in a soft artery under low tension; the dicrotic wave then becomes prominent and is appreciated by the finger as a second, weak pulsation: this is dicrotism of the pulse.

Dicrotism is found in febrile disease, especially in typhoid. In tracings, the dicrotic wave appears in various types according as its record appears above or below the base line: in a *predicrotic* pulse the elevation begins before the descending limb reaches the base line (moderate fever); in a *dicrotic* pulse, as the descending limb reaches the base line; in a *postdicrotic* pulse it begins below the base line (higher fever). In the tracing of a monocrotic pulse (very high fever), no dicrotism is noticeable.

Pulsus tardus shows a slowly rising ascending limb, a *rounded summit*, no tidal wave, no dicrotic wave (senile pulse). *Pulsus celer et altus* has a perpendicular ascending limb, no dicrotic elevation, several tidal waves.

The venous pulse shows a negative picture of the arterial pulse. The anacrotic limb is long drawn out and possesses a notch (*anadicrotic*), the katacrotic falls almost perpendicularly (*katamonocrotic*). The second limb of the tracing corresponds to the systole of the right auricle, the katamonocrotic limb to the ventricular contraction (auricular diastole). On the other hand, in *insuffi-*

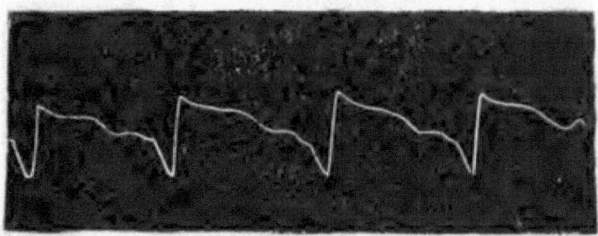

Fig. 30.—Sphygmographic Tracing of the Radial Artery of a Healthy Young Man.

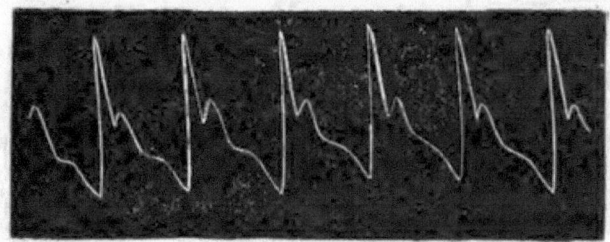

Fig. 31.—Sphygmographic Tracing of the Radial Artery in Aortic Insufficiency.

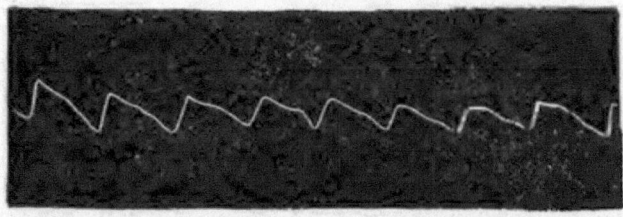

Fig. 32.—Sphygmographic Tracing of the Radial Artery in Aortic Stenosis.

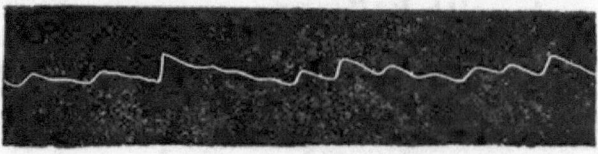

Fig. 33.—Sphygmographic Tracing of the Radial Artery in Mitral Stenosis.

ciency of the tricuspid, the expansion of the vein due to the systole of the auricle does not follow the systolic collapse of the heart, but there is *an additional expansion synchronous with the ventricular systole*. The collapse of the vein takes place in the next diastole. The "actual" venous pulse begins, then, in the ventricular diastole (auricular systole), lasts throughout the entire systole, and ends only at the beginning of the following diastole (cf. p. 164).

THE EXAMINATION OF THE URINE IN CARDIAC CASES

Since the excretion of the urine is dependent, in part, upon the arterial blood-pressure, a diminution of the arterial or increase of the venous blood-pressure is recognized in the *diminished* amount of urine. In cases of cardiac weakness or disturbed compensation the urine is scant, dark-red, of high specific gravity, with profuse brick-red sediment, frequently containing a trace of albumin.

In congestive conditions of long standing, a secondary nephritis may develop with hyaline and granular casts in the sediment; the nephritis of congestion may even lead to granular atrophy.

The improvement in the heart-disease shows itself plainly in the increased amount of urine and the disappearance of the albuminuria.

SYMPTOMS OF THE MOST IMPORTANT CARDIAC DISEASES

Common symptoms to all these diseases are: *in the stage of compensation, absence of any essential difficulties;* in the stage of disturbed compensation: cyanosis, dyspnœa, œdema, urine of congestion.

IDIOPATHIC CARDIAC DISEASES

Hypertrophy, or dilatation of the ventricle, with clear tones, or even the systolic murmur of relative insufficiency.

Prolonged over-activity of the left ventricle in arteriosclerosis, chronic *nephritis* and *contracted kidney*, over-exertion *for a long period*, over-indulgence in drink (beer heart),—all these factors aid in production of hypertrophy of the left ventricle with subsequent dilatation.

Cardiac weakness without a previous hypertrophy, partly with and partly without dilatation of the heart (*debilitas cordis*, weak heart), is developed (1) by the action of toxic substances upon the heart-muscle: *alcoholism* (whiskey drinking), excessive use of tobacco, *infectious toxines* (myocarditis after diphtheria, typhoid, etc.); (2) by improper nourishment: inanition, anæmia, senility.

Hypertrophy of the right ventricle, with subsequent dilatation, arises in all obstructions to the pulmonary circulation (emphysema, pleural thickenings, kypho-scoliosis).

A *fatty heart* is the heart which gives disagreeable phenomena to very stout people. Cardiac difficulties of the first degree are those produced by the deposit of fat in the abdomen as well as in the pericardium; difficulties of greater intensity are referable to the growth of fat into the heart-muscle. When the cardiac muscle undergoes fatty degeneration, troubles of the most serious nature are evoked, leading even to death. As fatty degeneration of the heart is mostly the ultimate outcome of fatty deposits about the heart, many writers think that the expression "fatty heart" should be reserved for this final stage, and a term expressive of "cardiac weakness" should be employed for the cardiac troubles of the obese.

Valvular Lesions

Valvular lesions are the result of valvular endocarditis, which is productive of contraction (insufficiency) or adhesions (stenosis) of the flaps of the valves. The endocarditis is usually a sequel of the acute infectious diseases, particularly of articular rheumatism. Besides this etiology, arteriosclerosis may act as a cause.

When the valvular lesion is completed, hypertrophy, and, later, dilatation ensues, in consequence of the forced increase of function on the part of the ventricle. In aortic lesions it is the left, in mitral lesions the right ventricle which is thus affected.

Aortic insufficiency (frequent). Marked cardiac impulse, apex-beat displaced to the left and downward, cardiac dulness extended to the left. Diastolic murmur in the direction of the blood-current regurgitating from the aorta. Best heard in the middle of the sternum and to the left of it in the 3d intercostal space. Frequently a systolic murmur at the apex and accentuated 2d pulmonic sound. *Pulsus celer et altus.* Strong pulsation of the carotids; systolic murmur in the carotids with absence of the 2d sound. Humming sound in the femoral. Capillary pulse.

Aortic stenosis (rarely alone). Weak cardiac impulse, apex-beat displaced to the left and downward, but much less than in insufficiency. Cardiac dulness extended to the left. Loud systolic murmur over the aorta (frequently *frémissement cataire*); heard much weaker over the other valves. Second aortic sound weak or absent. Pulse small, of low tension, slow. Frequently facial pallor; sometimes syncope, convulsions, attacks of vertigo.

Mitral insufficiency (most common valvular lesion). Cardiac impulse moderately strong, apex-beat frequently displaced to the left. Cardiac dulness extended to the right. At the apex and the pulmonic, a systolic murmur; 2d pulmonic sound much accentuated. Marked dilatation of the right ventricle is accompanied by congestion of the liver and venous pulse.

Mitral stenosis (rarely alone). Cardiac impulse quite marked, epigastric pulsation common. Cardiac dulness extended to the right. At the apex a presystolic murmur (frequently *frémissement cataire*), intensified, rough, systolic sound. Forcible closure of the pulmonic valve. Pulse slow and small, slightly irregular.

Tricuspid insufficiency. Cardiac dulness extended to the right, systolic murmur over the tricuspid valve, diminished intensity of 2d pulmonic sound, actual venous pulse, pulsation of the liver. The symptoms of this disease are rarely due to anatomical changes in the valve, but are more frequently to be traced to relative insufficiency when marked dilatation of the right ventricle exists as a common sequel of mitral lesions.

The *pulmonic* lesions are always congenital and are quite rare. They are marked by deep cyanosis, the cardiac dulness is extended to the right, and the murmurs heard correspond to the lesion present.

PERICARDITIS

Cardiac dulness increased in area in the form of an equilateral triangle, the apex of which points upward. The upper limit of cardiac dulness, 2d to 3d rib. Cardiac impulse and apex-beat weak or not palpable; increased in intensity when the patient bends forward, and are usually felt somewhat internal to the area of dulness. Heart-sounds very weak. Friction sounds not synchronous with the cardiac beat are common. Pain in the region of the heart not uncommon in swallowing.

Pericardial plates (*concretio pericardii*), after the healing of pericarditis, sometimes lead to the symptoms of cardiac weakness without offering percutory or auscultatory signs. In some cases the diagnosis is made possible by the systolic retraction of the region of the apex-beat, or by the presence of pulsus paradoxus.

SYMPTOMS OF THE PRINCIPAL DISEASES OF THE LARGE BLOOD-VESSELS

Arterio-sclerosis (atheroma of the arteries). — The palpable arteries (especially the radial and temporal) hard and tortuous, sometimes uneven and rough. Pulse under tension. mostly sluggish. Hypertrophy and dilatation of the left ventricle. Frequently a systolic murmur at the apex; less often a diastolic murmur over the aorta. Angina pectoris or cardiac asthma, common (coronary sclerosis).

Atheroma of the peripheral arteries does not always indicate atheroma of the large vessels and of the heart. On the other hand, a high degree of sclerosis of the aorta (endaortitis) may be present, without any apparent disease of the radial arteries.

Aneurysm of the thoracic aorta. — Dulness over the upper part of the sternum and its neighborhood. Over this area a thrilling systolic or diastolic murmur, audible and palpable. As the disease advances, the pulsating tumor appears to the left of the sternum in the region of the 2d or 3d rib. Frequently hypertrophy and dilatation of the left ventricle. Unevenness of the two radial pulses.

CHAPTER VIII

THE EXAMINATION OF THE URINE

By the examination of the urine we obtain information of :—

1. *The condition of the kidneys* and of the bladder. The normal renal epithelium does not permit the albumin in the blood to pass. In kidney disease, albumin and histological elements appear in the urine. Diseases of the bladder are associated with definite processes of urinary decomposition.

2. *The progress of the metabolic phenomena.* The end products of the disintegration of the albuminoids (urea, etc.) are voided in the urine; from an examination of this excretion we can determine the quantitative relations existing between the nitrogenous income and expenditure, which undergo specific changes in disorders of metabolism, as well as the admixture of certain definite compounds which result from certain anomalies of the metabolic processes, or which escape the chemical change or conversion which they should normally undergo (sugar, acetone, etc.).

3. *The force of the heart* (see p. 176).

4. *Diseases in other organs,* which permit certain substances to escape into the blood and from it into the urine. In disease of the liver, biliary coloring matter, in disease of the intestine, indican, and in suppuration, peptone, appear in the urine.

5. *The presence of heterogeneous compounds* which may be administered from without, such as iodine and mercury.

The quantity of urine passed in 24 hours amounts on the average to 1500 cc., and depends within further limits on the amount of fluids partaken of. A daily quantity below 500 cc. and above 3000 cc. is usually an indication of disease.

Diminution of the amount of urine occurs when there is profuse perspiration and in cases of diarrhœa, in fever, in cardiac weakness, in acute and often in chronic nephritis, in cases of effusions and transudations.

In cardiac diseases and in acute nephritis the prognosis as to immediate danger is for the most part made from the amount of urine passed. In effusions, *e.g.*, pleuritic, the first indication of the absorption of the fluid is shown by an increase of the secretion of urine. The excretion of a small amount of urine is often a sign of *deficient nutrition.*

Increase of the amount of urine is observed in diabetes mellitus and diabetes insipidus, in atrophic (small) kidney, in cases of effusions and transudations when absorption occurs, and often during the period of convalescence from an acute disease.

The specific gravity of the urine varies in health between 1010 and 1025, and is in inverse proportion to the amount excreted.

An unusually low specific gravity occurs in contracted kidney and in diabetes insipidus. A very high specific gravity is observed in diabetes mellitus, in fever where the amount passed is small, in wasting diseases, and in nephritis.

It is possible to determine from the specific gravity the amount of the solid constituents in 1000 cc. of the urine in this manner: multiply the last two ciphers of the specific gravity by 2.33 (*Häser's* coefficient). For example, should the urine to be examined have a specific gravity of 1015, then $15 \times 2.33 = 34.95$; hence the solid constituents in the urine are 34.95 in every 1000 cc.

The color of the urine. — The normal color of the urine is a more or less deep yellow; the smaller the quantity the deeper the color. It is of a *deep yellowish-red color* when the urine contains urobilin (see p. 192), and it is *red (like beef solution)* when it contains *blood* (see p. 190). A *brown color* with a yellow foam indicates the presence of *biliary coloring matter* (see p. 191). In *carbolic acid*, seldom in salicylic acid, *intoxication* the color varies from an *olive-green* to a *black*. It is *yellowish-green* after the use of *rhubarb* and *santonin*. The color grows darker when exposed to the air, either because melanin is present or because carbolic acid was taken (see p. 202).

The color directs our attention to the abnormal elements which the urine contains, whereupon these may be detected by a chemical and microscopical examination.

Cloudy urine. — Normal urine is clear. The significance of cloudy urine depends on its *reaction*, which may be tested by litmus and by the smell (see below).

Cloudiness of *acid* urine is owing either to the urates which may be present, and in this case the urine becomes clear on heating in a test-tube, or to *organic histological elements* (see below) when the urine will not become clear on heating. These anatomical elements may be detected by the use of the microscope.

The organic elements consist of casts, renal epithelium, or pus cells. Pus is detected by boiling with potassic hydrate, which produces a mucilaginous mixture.

Cloudiness of *alkaline* urine arises from the presence of phosphates, seldom from oxalate of lime, or from organic elements which may be determined by the use of the microscope.

On the addition of an acid the cloudiness arising from the presence of salts will disappear, whereas it will naturally not be dispelled when organic elements give rise to it. Pus is determined here also by the mucilaginous appearance of the urine when boiled with potassic hydrate.

The reaction. — Normal urine is acid. Should the urine to be examined be alkaline, we must determine whether this reaction be due to the presence of a fixed alkali (calcic carbonate) or to a volatile one (ammonium carbonate). This is determined by holding over the urine a piece of moistened red litmus paper. Should this paper turn blue without agitating the urine, then the alkalinity is due to the ammonium carbonate in the urine; should, on the other hand, the blue color result only after dipping the paper into the urine, the alkalinity is due to the presence of *potassium* carbonate or *sodium* carbonate. If we hold a glass rod moistened with hydrochloric acid over a dish containing ammoniacal urine, a white cloud of ammonium chloride, NH_4Cl (sal ammoniac), will be produced.

The reaction of the urine depends upon the quantitative relations existing between the acids and bases. Urine contains the following acids: hydrochloric, sulphuric, phosphoric, uric, kreatin, a little hippuric and oxalic; the following bases are present: potassium, sodium, calcium, magnesium, ammonium, kreatinin, xanthin, and hypoxanthin.

By the disintegration (metabolism) of the albuminoids, of the lecithins, and of nuclein, sulphuric acid and phosphoric acid are copiously produced; it is on this account that after the ingestion of meat, cheese, legumens, and cereals, the urine becomes strongly acid.

While *hydrochloric acid* is forming compounds in the stomach, in the first act of the digestion of a meal rich in albuminoids the acid reaction of the urine is diminished; should the hydrochloric acid be removed by vomiting or by washing out the stomach, the urine must likewise be less acid and more alkaline.

A direct increase of alkalinity occurs in the administration of those *potassium salts whose acids are easily oxidized*, potassium tartrate, citrate, and malate, which are oxidized into potassium carbonate. These salts are abundantly present in fruits, berries, and in potatoes. The eating of large quantities of fruit will, therefore, render the urine alkaline, as will naturally the medicinal administration of sodium carbonate.

The alkaline reaction of the urine from presence of a fixed alkali appears soon after a hearty meal; after a hearty indulgence in fruits, berries, or potatoes; in diseases of the stomach where vomiting occurs or where the stomach has been washed out; in cases of absorption of alkaline effusions; after the use of alkaline waters and medicines.

The alkaline reaction of the urine from the presence of a volatile alkali is produced by the development of bacteria in the urine, which decompose the urea into ammonium carbonate, thus $(CONH_2)_2 + 2 H_2O = CO(NH_4O)_2$. This decomposition takes place in all urine that has stood for a long time, especially in a warm place. Should the urine be decomposed as soon as it is voided from the bladder, there is a *cystitis* present.

Ammoniacal decomposition produces a very characteristic odor, which is easily recognized.

Every alkaline urine contains a sediment composed of ammonio-magnesium phosphates (triple phosphates), calcium phosphate, and often of calcium carbonate. In decomposition the triple phosphates are most prominent, and at the same time the urine is rich in bacteria. When the alkalinity is due to a fixed alkali, the sediment contains much calcium phosphate and calcium carbonate.

THE CHEMICAL EXAMINATION OF THE PATHOLOGICAL ELEMENTS OF THE URINE

Albumin. — The presence of albumin (serum albumin and serum globulin) in the urine indicates a lesion of the renal epithelium. Under this head it may deal with mild or intense disturbances of nutrition (a congestion, an anæmia, cloudy swelling, fatty metamorphosis). Pronounced persistent albuminuria indicates nephritis.

In all urine of human beings nominal traces of albumin may be found, still these can only be detected by especially fine methods of analysis (*normal albuminuria*). Small quantities of albumin appear in some persons as a *transient* constituent of the urine after a meal rich in albuminoids, after intense physical strains, after hot baths, and after psychical excitement. It is then known as *physiological albuminuria*.

If the urine contain *blood* or *pus*, the albumin derived from their plasma is dissolved by the urine and even after filtration the urine will show the reaction for albumin; such an albuminuria does not denote in itself a disease of the kidneys. It is known as *spurious albuminuria*.

The examination of a patient is incomplete without an examination of the urine for albumin. When a small quantity is detected, the urine should be again examined from specimens passed at different periods of the day (the urine passed on awakening in the morning is often free from albumin even when the kidneys are diseased).

Of practical significance is *intermittent albuminuria*, which is the condition obtaining when small quantities of albumin appear *transiently* in the urine at different times and mostly in otherwise healthy young persons. Even if the suspicion of nephritis must be entertained, the progress of the disorder is most favorable. *Cyclical albuminuria* occupies a peculiar position; it consists in this: the urine in the early morning is always free of albumin, but shows, if examined every two hours, varying amounts of albumin, which reach the maximum mostly at night. Though in these cases the question of a chronic nephritis can not be excluded with absolute certainty, still a more frequent disappearance of albumin is observed.

Tests for Albumin

Freshly voided urine should be used and preferably the last portion of the urine which has been passed.

For the preparation of the tests the urine should be *clear*, it should be *filtered*.

1. *The boiling test.* The urine is placed in a test-tube to one-quarter of its length and is then heated in the flame until it boils, when $\frac{1}{10}$ volume of nitric acid is added. Should the urine have become cloudy and the cloud have become cleared up during the heating, the cloudiness was due to the presence of acid sodium urate. Should the cloudiness exist on heating and become dispelled only after the addition of nitric acid, it was due to the presence of either calcium carbonate or calcium phosphate. Persistency of the cloudiness or the appearance of a precipitate after the addition of the nitric acid indicates albumin.

Should the portion tested in this manner be permitted to stand in the test-tube over night, the precipitate will have gravitated to the bottom of the tube. The volume of the precipitate in its relation to the volume of the original fluid tested will give an approximate quantitative test. This relation is as follows: slight cloudiness of the fluid (trace of albumin) corresponds to 0.01 per cent.; when the precipitate occupies merely the bottom of the tube, there is 0.05 per cent. of albumin; when the precipitate is

$\frac{1}{10}$ of the volume of the tube contents, there is 0.1 per cent. of albumin; when $\frac{1}{4}$ of the volume, 0.25 per cent.; $\frac{1}{2}$ of the volume, 0.5 per cent.; $\frac{3}{4}$ of the volume, 1 per cent.; when the whole contents become solid, between 2 and 3 per cent. are present.

2. *Test* with *acetic acid* and *potassium ferrocyanide*. To the cold urine in a test-tube a few drops of acetic acid,[1] then some of a 5 per cent. solution of potassium ferrocyanide is added drop by drop; if albumin be present there will be immediately or in a few moments a flaky precipitation of the coagulated albumin.

3. *Heller's test*. Concentrated nitric acid is slowly added to a small quantity of urine contained in a test-tube. The tube should be held inclined that the acid may flow down the wall of the tube. The acid will sink to the bottom and at the dividing line between acid and urine, if albumin be present, a narrow cloudy white ring will be formed. Still a similar ring may be produced in the presence of uric acid, urea nitrate, and of the resinous acids which result from the administration of resinous substances like turpentine, copaiba, etc. The ring formed from the presence of the resinous acids is soluble in alcohol, that from urea forms only after long standing and is also distinctly crystalline; the uric acid ring is not as sharply defined as is the ring arising from the presence of albumin, it rises higher in the urine and is only produced in very concentrated urine.

Quantitative determination of the albumin. For clinical purposes a sufficiently accurate quantitative analysis may be made by means of *Esbach's* albuminometer. The albuminometer is filled to the line marked U with the urine to be examined, the Esbach fluid, which consists of citric acid 5, picronitric acid 2.5, distilled water 245, is then added to the line marked R, the mixture is then shaken, and on the following day the deposit which has collected will reach a line at the lower portion of the tube, which is graduated, denoting in grammes the amount of albumin present in 1000 cc. of urine.

For scientific purposes, *weighing* of the precipitated albumin is the best method. This method consists of placing 100 cc. of urine in a porcelain dish, adding sufficient acetic acid to render the urine slightly acid, and then heating the contents of the dish to the boiling-point. The fluid containing the precipitate is then filtered, the filter having previously been thoroughly dried and weighed. The precipitate is well washed with hot water, then with alcohol

---

[1] Sometimes a cloudiness develops on the addition of the acetic acid; this is due to mucin or nucleo-albumin, and should be removed by filtration.

and ether; it is dried and weighed; the additional weight will represent the percentage amount of albumin.

In addition to the albumin precipitated by boiling, the urine may contain forms of albumin which are *not precipitated by heat*. These forms also result in gastric digestion of albumin. It has recently been proven that the end product of the gastric digestion of albumin (*Kühne's peptone*) does not appear in the urine, but that only premature stages of the same (albumoses and propeptones) are there found. It is a mistake, therefore, to speak of a *peptonuria;* we ought to call that condition *albumosuria*. However, this term has up to this time been very little used.

The albumoses (peptone in the old sense) are not precipitated by heat and by acids; their presence is determined by the biuret test after the other albuminous bodies have been precipitated and removed by filtration.

The albumoses are found in urine, especially in cases of absorption of *purulent* or fibrinous exudations (*pyogenic albumosuria*), especially in cases of pneumonia shortly before and after the crisis, in purulent meningitis, in peritonitis, in empyema, at times in cases of intestinal ulceration, in many diseases of the liver, in puerperal fever (*enterogenous, hepatogenous,* and *puerperal albumosuria*). Albumosuria is chiefly diagnostic of suppuration, but should be estimated only with great precaution.

Test for peptone (*Hofmeister's test*). Add to 500 cc. of urine 50 cc. of a concentrated solution of sodium acetate; drop slowly into this a concentrated solution of ferric chloride, until the resulting fluid remains at a red color. At this stage carefully add a solution of potassium hydrate, until the previously strongly acid mixture becomes neutral, or is only slightly acid; then boil and filter after the fluid becomes cool. Should the filtered solution be found by the acetic acid and potassium ferrocyanide test to be free from albumin, the *biuret test* is then used: a few drops of potassium hydrate and a few of a 1 per cent. solution of cupric sulphate being added, a beautiful red color results if peptone be present.

The recently proposed method of *Salkowski* is much simpler and just as sensitive: 20 to 50 cc. of urine which has been freed from albumin, and to which some hydrochloric acid has been added, is precipitated by phospho-molybdic acid; the precipitate is heated, washed with water, finally dissolved in a weak solution of sodium hydrate, and again heated until it assumes a yellow color. After cooling, the biuret test is put into use (see above).

Blood. — We recognize the presence of blood by the color of the urine. This is bright red with a greenish shimmer (like meat solution) in the presence of *oxyhæmoglobin;* dirty brownish-red in the presence of *methæmoglobin*. Nevertheless, we ought not decide that blood is present from the color alone; a *microscopical* examination of the sediment (see p. 210) and the chemical tests for blood should also be made. A spectroscopic examination is only seldom used (Chap. XI.).

Heller's test. To the urine in a test-tube ¼ of its volume of potassium hydrate solution is added, and the mixture is boiled. After a short time the earthy phosphates precipitate (magnesium and calcium phosphates). When blood is present the flaky precipitates will be of a reddish-brown color (in normal urine they are grayish-white). The color is best determined after the precipitate has fully collected together.

Van Deen's test (guaiac test). Add to the urine 2 cc. of tincture of guaiac and 2 cc. of old oil of turpentine, then shake the mixture thoroughly; if blood be present, the whole solution will become blue in a short time (pus will give the same reaction). Instead of old turpentine oil the following mixture may be used (*Huhnerfeld's test*): glacial acetic acid, 2; distilled water, 2; oil of turpentine, 100; absolute alcohol, 100; chloroform, 100.

Blood in the urine signifies a *hæmaturia* or a *hæmoglobinuria*. *Hæmaturia* indicates the presence of the *red blood corpuscles* in the urine; the blood may come from the kidney, the pelvis of the kidney, the bladder, or the urethra. The causes of hæmaturia are acute nephritis, renal calculi, hæmorrhagic infarct of the kidneys, tumors of the kidneys, renal hæmophilia, pyelonephritis, acute cystitis, carcinoma of the bladder, and *stone* in the *bladder*.

Temporary hæmaturia may also be the result of physical over-exertion.

In making a differential diagnosis the following data should be considered: the urine is small in amount in acute nephritis and contains more albumin than the amount of blood it contains would justify; in addition to these facts there are casts in the deposits, and the patient, as a rule, has dropsy. Hæmorrhagic renal infarction is associated with cardiac disease, and is accompanied by fever and pains, the blood disappearing from the urine in a few days. The hæmorrhage from renal calculi is preceded by attacks of colic, and the calculus may be passed with the urine. In pyelonephritis a purulent deposit will also be present in the urine. In cystitis there is pain in the bladder and vesicular tenesmus, while the urine contains pus at the same time. In cases of tumor it is sometimes possible to find small particles of the growth in the urine.

When a careful consideration of all the causes mentioned fails to discover the origin of the hæmaturia, it may be possible that, as occurs in rare instances, the hæmorrhage arises from a healthy kidney, of which two varieties are known.

1. The *hæmaturia of bleeders* (renal hæmophilia) may be diagnosticated when the patient's family is one of bleeders, and when he himself has given previous evidence of hæmophiliac dyscrasia. 2. *Angioneurotic hæmaturia*, which occurs in neuropathic individuals; in which case one may succeed in stopping the hæmorrhage by suggestion or hydriatic measures. Hæmorrhage from a normal kidney should only be considered when every other diagnostic possibility has been excluded.

In doubtful cases, severe pain in the lumbar region may indicate the origin of blood as coming from the *kidney;* proof of this is the presence of red blood corpuscles in the urinary sediment, when they occur as blood casts.

We can occasionally determine whether the hæmorrhage comes from the bladder by catheterizing and washing out the bladder after urination has occurred. If the blood comes from the bladder, the returning water will be blood-stained. In spite of this aid it is at times extremely difficult to determine whether the source of the blood is from the kidney or bladder. In such cases the source of the bleeding may be determined by the *cystoscope* in expert hands. The cystoscope will also determine from which kidney the blood arises.

Hæmoglobinuria is the appearance of a solution of the coloring matter (*without* the corpuscles) of the blood in the urine. It results in consequence of the red blood corpuscles being dissolved from various causes, namely, from poisons (potassium chlorate, mussel poisoning, etc.); it appears after transfusion of blood, it results from burns and as a specific disease (periodical hæmoglobinuria), and often after exposure.

Biliary coloring matter appears in the urine either reduced as *hydrobilirubin*, the same as *urobilin*, or unreduced in the form of *bilirubin*. The special biliary coloring matter is detected by the beer-brown color of the urine and the yellow color of the foam which is produced by shaking. Urine containing urobilin is yellowish-red with a tinge of orange. For a positive determination special chemical and spectroscopic analyses are required.

Gmelin's test for bilirubin. Concentrated hydrochloric acid is mixed in a test-tube with one or two drops of fuming nitric acid. The urine is then carefully poured down the walls of the tube; the acid and the urine form two distinct layers, and at their junction a colored ring will be formed; this ring is first green (*biliverdin*), then violet, then red, then yellow (*choletelin*), and finally it assumes a dirty color. The production of a blue ring arises from the presence of *indican.*

The same diagnostic conclusions should be formed when bilirubin is shown to be present in the urine as would from the presence of icterus.

Test for urobilin. Render the urine alkaline by adding some ammonia to the urine in a test-tube, then add 8–10 drops of a 10 per cent. solution of zinc chlorate and filter quickly. View the filtered solution against a dark background, when it will appear green, but by transmitted light it will be rose-red. *Spectroscopical* examination of the urine will detect the smallest portion by showing an absorption line between the green and blue fields.

Hydrobilirubin (urobilin) occurs in many cases of icterus in which the color of the skin of the patient is a dirty yellow (urobilin icterus), in states of passive congestion, and in high fever. In addition it occurs where large extravasations of blood have been absorbed (as a result of the reduction of the coloring matter of the blood identical with bilirubin). With a sufficient regard for the other causes a diagnosis of internal hæmorrhage (apoplexy, infarct, etc.) may be justified from the presence of a large amount of urobilin in the urine.

Sugar. — Urine which contains sugar is usually passed in large amounts; it is of light color, and of a high specific gravity.

The proof of sugar in the urine depends upon its following characters: —

1. Glucose is colored brown (caramel formation) when boiled with potassium hydrate.
2. Glucose has the property to reduce other bodies under high temperatures.
3. Glucose ferments in the presence of yeast and is converted into alcohol and carbonic acid ($C_6H_{12}O_6 = 2\,C_2H_5OH + 2\,CO_2$).
4. Glucose turns the plane of polarized light to the right.
5. Glucose unites with phenyl hydrazin into a characteristic chemical compound (glucosazon).

Qualitative Analysis for Sugar

1. *Moore's test* consists in adding to the urine in a test-tube ⅓ of its volume of potassium hydrate solution and boiling the mixture several times in succession; in the presence of sugar (glucose) the urine will assume a *brown color*.

2. *Reduction tests.* a. *Trommer's test* consists in adding to the urine ⅓ of its volume of potassium hydrate solution and again adding enough of a 10 per cent. solution of cupric sulphate until the blue precipitate which is formed thereby is entirely dissolved while agitating the mixture; as soon as the addition of another drop of the copper solution is precipitated and is not dissolved by agitating the contents of the tube, stop the addition of the cupric sulphate and heat carefully the upper part of the solution in the tube by holding that part of the tube over a flame. If sugar be present, a yellowish-red precipitate will be formed as soon as that part of the mixture rises to 60° to 70° C. before the boiling-point is reached. As soon as the precipitate forms, stop the heating.

The reaction proceeds as follows: $CuSO_4 + 2 KHO = Cu(HO)_2$ (cupric hydroxide) $+ K_2SO_4$. Cupric hydroxide, $Cu(HO)_2$, in itself may be converted by heat into black cupric oxide, $(CuO) + H_2O$. The 2 CuO, or the $Cu(OH)_2$, gives up by the heat one atom of O to the sugar, thus producing Cu_2O, yellow copper oxydul (cupro-oxyd), or CuOH, brown copper oxydulhydrate (cuprohydoxyd) respectively.

The ordinary dissolving of the cupric hydroxide with the production of an *azure*-blue color does not determine the presence of sugar; for the same reaction may be produced when albumin, ammonia, and other organic substances are present.

Even the change of color of the tested solution to yellow *without the precipitation* of a yellow precipitate does not prove the existence of sugar; for uric acid and kreatinin will likewise reduce the oxide of copper, but retain copper oxydul (cupro-oxyd) in solution.

In the strict acceptation the production of copper oxydul does not prove the existence of sugar, but *only that a reducing substance* is present in the urine. Other reducing substances may occur in the urine especially after the ingestion of certain substances (chloral hydrate, camphor, chloroform, turpentine, benzoic acid, salicylic acid, copaiba and cubebs). When any of these have been taken, in addition to *Trommer's* test, a control analysis for sugar should be made by fermentation and the polariscope.

Trommer's test may be made more quickly by adding slowly an equal volume of a doubly diluted *Fehling's* solution to a few cubic centimetres of the boiling urine. If sugar be present, a yellow precipitate will be thrown down. The Fehling's solution, however, should be previously tested to see if it itself will not form a similar precipitate on boiling.

    b. Böttcher's test. Saturate 10 cc. of urine with some finely powdered sodium carbonate, add thereto a very small quantity of *basic bismuth nitrate*, and boil for some minutes. The production of a *black color* in the urine indicates the presence of sugar; only when organic substances (albumin, mucus, pus, and blood) are present in the urine may this production of bismuth sulphide lead to error. A very convenient method for conducting this test is by the use of *Nylander's solution*, which consists of Rochelle salts (potassium sodium nitrate, 4; 10 per cent. solution of sodium hydrate, 100; bismuth subnitrate, 2; heated together and then filtered). If urine to which $\frac{1}{10}$ of its volume of *Nylander's* solution has been added be boiled, should sugar be present a brown or black color will be produced.

    c. Rubner's test. Add to the urine a knife tip full of plumbic acetate; a precipitate of plumbic phosphate and plumbic sulphate will be formed. Remove the precipitate by filtering and to the filtered solution add some ammonia, whereupon a white precipitate of plumbic oxide will be formed. On boiling at this point of the procedure the white precipitate will become rose-red in the presence of sugar, owing to the formation of a higher oxide (red oxide of lead, Pb_3O_4).

    3. *Fermentation test.* This test is one of the most reliable of all tests for sugar. It is performed as follows: shake the test-tube containing the urine to which a small cube of fresh compressd yeast has been added, then pour this mixture into a *fermentation tube*; turn the tube so that the long arm will be filled completely by the urine. When the apparatus has been filled, pour some mercury into it in order to close the long arm, then place the entire apparatus in a warm place at about 24° C. Should the urine contain sugar, gas bubbles composed of CO_2 will arise in the tube in a few hours. That this gas is truly CO_2 may be proven by the addition of potassium hydrate, which will quickly absorb the gas. For a control of this test, two more fermentation tubes should be prepared, one with a mixture of a solution of glucose and yeast (to determine that the yeast is efficient), the other with normal urine and yeast (this should not generate gas, and should show that the yeast itself is free from sugar).

A fermentation tube may be improvised from a test-tube, as may be seen from Fig. 34. The picture shows a fermentation tube made after *Moritz's* suggestion and needs no further explanation. *Pavy's* fermentation tube may be prepared by introducing a glass tube through the stopper of a test-tube which has been completely filled with the urine to be examined; the tube should reach almost to the bottom of the test-tube. The outer end of the tube should be bent and placed in a vessel containing water; the gas which forms collects under the stopper, and by pressure the urine rises in the tube to escape into the vessel containing the water.

4. *Phenyl hydrazin test* is carried out as follows: to 10 cc. of a heated 10 per cent. solution of sodium acetate, as much acid phenyl hydrazin is added as will go on the end of a knife; to this mixture 10 cc. of urine are added, and the test-tube containing the entire solution is placed in a water bath at 100° C. for an hour. When sugar is present, numerous yellow crystals of phenyl glucosazon are formed and precipitated. The latter compound melts at a temperature of 205° C. Should the urine contain but little sugar, the solution is centrifuged, and the sediment examined microscopically for the crystals.

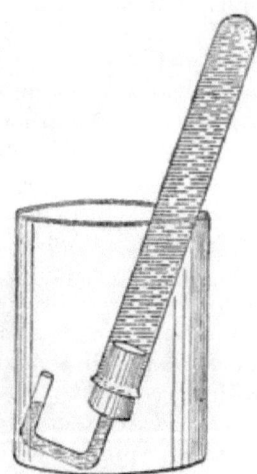

FIG. 34. — IMPROVISED FERMENTATION APPARATUS. (MORITZ.)

Sugar may appear in the urine in health after the ingestion of any meal containing more than 150 g. of glucose (alimentary or physiological glycosuria). In general a suspicion of a diabetic tendency should be held in every case of glycosuria, even when sugar in the urine is transient and only small in amount.

Alimentary glycosuria has sometimes been observed after the ingestion of moderate amounts of glucose (about 100 g.) in the obese, in drunkards, in cases of *Basedow's* disease, as well as in traumatic neurosis, and also in those who have used thyreoid preparations. In excessive diuresis a small quantity of sugar may appear in the urine.

The persistent or long-continued appearance of more than ½ per cent. of sugar in the urine occurs only in *diabetes mellitus*. A complete diagnosis of this disease demands the quantitative determination of the amount of sugar passed (see Chap. X., Diseases of Metabolism).

Quantitative Analysis of Sugar in the Urine

Estimation of the amount of sugar by Moore's test. Urine heated with potassium hydrate is of a straw-yellow color when it contains about 1 per cent. of sugar, amber-yellow with 2 per cent., brown, like Jamaica rum, with 5 per cent., and dark brown with 7 per cent. This estimation is but little reliable, and will give to the one who is expert in its use only approximate results.

Estimation of the amount with Fehling's solution. Fehling's solution consists of 34.639 g. of crystallized cupric sulphate, 173 of Rochelle salts, 100 c.c. of officinal sodium hydrate solution, and water to make the entire mixture 1000 c.c.; 1 c.c. of this solution is reduced by 5 mg. of glucose.

Two cubic centimetres of *Fehling's* solution are put into a test-tube containing 20 c.c. of water, by means of a pipette. The cupric oxide contained in this amount is reduced by exactly 1 cg. of sugar. Bring the contents of the test-tube to a boil; then add *drop by drop* the urine to be examined, and see by transmitted light after every drop whether the solution has become *decolorized*. As soon as it has, stop adding the urine. The number of drops of urine which have been added to produce decolorization will then have contained 0.01 g. of sugar. Twenty drops are regarded as equal to 1 c.c. of urine.

The following table gives the approximate relation of drops to percentage of sugar.

Drops	Percentage	Drops	Percentage
100	0.20	14	1.40
90	0.21	13	1.50
80	0.25	12	1.60
70	0.28	11	1.80
60	0.30	10	2.00
50	0.40	9	2.20
40	0.50	8	2.50
30 / 25	0.60	7	2.80
		6	3.30
20	1.00	5	4.00
19	1.05	4	5.00
18	1.10	3	6.60
17	1.15	2	10.00
16	1.20	1	20.00
15	1.30		

The results of these values are pretty reliable when the method is properly used.

Titration with the use of Fehling's solution. Strongly dilute 20 c.c. of *Fehling's* solution in a porcelain dish; dilute the urine ten times and place it in a graduated burette. Drop the urine slowly into the boiling *Fehling's* solution until all the copper has been precipitated and the solution has become entirely decolorized. The amount of sugar in the urine necessary to perform this reduction is 0.1 g.; hence the amount of urine used contains 0.1 g. of sugar. From this the percentage of sugar is easily reckoned. For instance, if 27 c.c. of the ten times diluted urine were used to reduce 20 c.c. of *Fehling's* solution, then 0.1 g. of sugar was contained in the 27 c.c. of the ten times diluted urine; therefore in 100 c.c. of the same dilution of urine there would be $\frac{0.1 \times 100}{27} = 0.37$; but since the urine was diluted ten times, the total percentage would be $10 \times 0.37 = 3.7$ per cent. In order to accurately judge the time to stop adding the urine, a small amount of the *Fehling's* solution that has been used is filtered and added to some potassium ferrocyanide solution; should some unreduced copper be still held in the solution, a brown color (cupric ferrocyanide) will be produced. The titration method is entirely accurate, but may lead to error on account of the other reducing substances (see p. 193) contained in the urine.

Estimation by Einhorn's saccharimeter. This instrument is the ordinary fermentation tube, whose long arm is empirically graduated. 10 c.c. of urine are mixed in a test-tube with a piece of compressed yeast about the size of a bean; the mixture is well shaken and poured into the fermentation tube. According to its specific gravity, the urine should be previously diluted; a specific gravity of 1018 to 1022 necessitates a double dilution, 1022 to 1028 a fivefold dilution, 1028 to 1038 a tenfold dilution. This method gives a pretty good result for urine containing less than 1 per cent. of sugar, but it is not always accurate in estimating the percentage of sugar in urine containing more than 1 per cent. However, it is quite serviceable for practical purposes, especially for continued estimations.

Determination of the specific gravity before and after fermentation. The temperature and the specific gravity of the urine are taken; then 100 to 200 c.c. of urine are placed in a flask containing some compressed yeast, and the mixture is allowed to stand at 24° C. After 24 hours, when the formation of gas has ceased, the mixture is filtered, cooled to the original temperature, and the specific gravity is then taken. (The urinometer should be standardized.)

Every degree less in specific gravity will indicate 0.23 per cent. of sugar, so that, if the specific gravity of the urine was originally 1032, and after fermentation 1022, the amount of sugar contained would be $10 \times 0.23 = 2.3$ per cent. *This method gives very accurate results* when the amount of sugar in the urine is over 0.5 per cent.

Estimation by polarization. 20 to 30 c.c. of urine are placed in a glass beaker, and a small quantity of powdered plumbic acetate is added; the resulting precipitate is then removed by filtration. The filtered solution is then placed in the glass tube of the polariscope (usually will contain 10 or 20 c.c. of fluid), special care being taken to free the tube of all air-bubbles. The interior of the tube should first be washed with some of the filtered mixture. From the scale attached to the polariscope, better from the average of three different readings of the scale, the percentage of sugar is determined in this way. The deviation to which the tube is directed is read off from the scale as degrees; this number is multiplied by 100, and the result divided by 53.1. If the tube of the polariscope is 20 cm. long, the quotient is divided by 2. (The description and theory of the polariscope may be found in the larger text-books.) Should the urine contain *albumin*, it should be precipitated before the polariscope is used, because albumin deviates light to the left.

The polarization of urine will give other results from the titration method: (1) if the urine contains other reducing substances besides sugar; (2) if it contains other substances which deviate light to the left, such as oxybutyric acid, which is found in severe cases of diabetes. In such cases it is wise to titrate and to polarize again after the proposed fermentation. Excepting these sources of error, the polarization gives accurate results.

Acetone and aceto-acetic acid. — Both these compounds appear abundantly in the urine after an extensive disintegration of the albuminoids of the body, especially in cases of high fever, in severe anæmias, and in some cases of carcinoma, in hasty consumption, in severe forms of diabetes, and in inanition. In severe diabetes acetonuria occurs even when there has been a total abstinence from the carbohydrates, and when tissue disintegration has not occurred. In addition acetone is found in the urine in cases of disturbances of digestion and in intestinal diseases. In the latter cases acetone has been found even in the contents of the intestine.

Aceto-acetic acid (CH_3COCH_2COOH) is tested by *Gerhardt's* method of reaction to *ferric chloride* solution. On the addition of Fe_2Cl_6 to the urine, a gray precipitate of iron phosphate is thrown down; on the further addition of the ferric chloride, if aceto-acetic acid be present, a deep *claret* color is produced. If the contents of the tube are now shaken, the resulting froth will have a red-violet tinge. On the addition of sulphuric acid the red color will disappear.

On boiling the urine the aceto-acetic acid will be converted into *acetone* and carbonic acid: —

$$CH_3COCH_2COOH = CH_3COCH_3 + CO_2.$$

The acetone is distilled over (*distil* about ½ litre of urine with a few drops of hydrochloric acid) and may then be tested by *Lieben's* test, which is conducted as follows: some of the distilled fluid is mixed in a test-tube with a few drops of a solution of iodine in potassium iodide (iodine, 2; potassium iodide, 10; distilled water, 200) and of potassium hydrate solution. Should acetone be present in the urine, an immediate precipitation of yellowish-white characteristically smelling iodoform will result.

Acetone is proved to exist in the urine by means of *Legal's* test, A few drops of a sodium nitro-cyanide solution are added to the urine to be tested and the mixture is then rendered strongly alkaline. A purple-red color changing gradually to a yellow will result; 2 to 3 drops of acetic acid are then added, and, if acetone be present, a color varying from carmine to purplish-red will form where the acid touches the solution. Still, it is doubtful whether acetone occurs preformed in the urine.

The claret-red ferric chloride reaction is of practical diagnostic importance, especially in *diabetes;* the long-continued presence of a large amount of acetone indicates a grave form of the disorder and should suggest a very unfavorable prognosis.

Ehrlich's diazo-reaction. — In many diseases there appear in the urine certain aromatic compounds of whose composition little is known, which unite with *sulpho-diazobenzol* in such a manner as to produce certain characteristic colors.

The chemical course of the *diazo-reaction:* sulphanilic acid ($C_6H_4NH_2SO_3H$), when united with nitrous acid (HNO_2), produces sulpho-diazo-benzol ($C_6H_5NNSO_4H$ [diazo = 2 atoms of N]). The latter body unites with many aromatic amido-compounds to produce colors. In order to produce sulpho-diazo-benzol in a fresh state at any moment, it is necessary to keep a solution of sulphanilic acid in hydrochloric acid, and when it is needed to carry out the test, to add to this solution a solution of sodium nitrite, whereby nitrous acid is formed in a free state which from the sulphanilic acid produces sulpho-diazo-benzol.

Method of conducting the diazo-reaction. Two solutions should be prepared and separately kept, viz.:—

1. Sulphanilic acid,   5.0
   Hydrochloric acid pure, 50.0
   Distilled water,   1000.0

2. Sodium nitrite,   0.5
   Distilled water, 100.0

In order to carry out the test, 50 c.c. of the sulphanilic acid solution and 1 c.c. of the sodium nitrite solution are mixed in a glass graduate. This mixture is added to the urine which is to be examined, in the proportion of half urine and half mixture, and to these is added an amount of ammonia equal to ⅓ of the volume of the combined urine and mixture. The entire mixture is then placed in a test-tube and violently shaken. If the *resulting froth* should assume a *deep red* color, the diazo-reaction is positively demonstrated.

The diazo-reaction is still more beautifully shown if we use *para-amido-aceto-phenon* in the proportion of 0.5 to 1000 instead of sulphanilic acid. This constitutes *Friedenwald's* modification of *Ehrlich's* reaction.

The diazo-reaction is obtained in the urine of cases of typhoid fever, pneumonia, measles, miliary tuberculosis, sepsis, and in severe cases of phthisis. It is absent in meningitis. The principal value of the diazo-reaction is its presence in cases of *typhoid fever*, where in doubtful cases it often decides the diagnosis and renders the relapsing character of the subsequent fever positive. The disappearance of the reaction shows that the infection has subsided. In phthisis the appearance of the reaction indicates a bad prognosis.

Fat in the urine is recognized by the milk-like cloudiness of the entire urine, which disappears when some potassium hydrate solution is added to it and the mixture well shaken with ether. Fat is often held in solution in the urine. On microscopical examination numerous small fat globules may be seen. Even on centrifuging the urine the fat may not in all cases be separated. This symptom is called *chyluria;* it represents an individual disease, common in the tropics, and often produced by the *filaria sanguinis* (see Chap. XII.). Chyluria is only seldom the result of occlusion of the thoracic duct; in many cases its etiology is obscure.

Melanin is the coloring matter of melanotic carcinoma, which at times is swept into the circulation and thus appears in the urine, rendering the urine in rare cases a dark black color. In other likewise rare cases an antecedent stage (*melanogen*) of the same coloring matter appears in the urine, from which the latter may be precipitated by ferric chloride.

Sulphuretted hydrogen occurs in rare cases in the urine of patients with cystitis, and is due to the action of certain bacteria which induce a peculiar urinary metamorphosis. H_2S may be recognized by its smell, like that of rotten eggs, or by holding over the urine some paper saturated with plumbic acetate, which will turn brown owing to the formation of plumbic sulphide.

CHEMICAL EXAMINATION OF THE NORMAL ELEMENTS OF THE URINE WHICH ARE QUANTITATIVELY CHANGED BY DISEASE

1. Inorganic Elements

Chlorides occur in the urine chiefly in the shape of sodium chloride; the normal amount depends upon the amount and character of food taken, but averages between 10 to 15 g. of NaCl. They are diminished in fever, especially in *pneumonia* and in inanition.

The test for chlorides is made in this manner: a few drops of nitric acid and a few of a 10 per cent. solution of silver nitrate are added to the urine; a cheesy precipitate will ordinarily be formed, but in the urine of cases of pneumonia, etc., there will often be only a slight cloudiness. Exact quantitative analysis is made by titration.

Phosphates occur in the urine as potassium sodium, calcium and magnesium phosphates. The quantity daily passed in the urine varies according to the food within wide limits, and is about 3 g. The quantitative analysis for phosphates is by methods of titration and has no essential diagnostic bearing.

Sulphates appear partly as potassium sulphate (preformed sulphuric acid), partly as phenol, indoxyl, and scatoxyl compounds (sulphuric ether compounds).

The separation of these two classes of compounds is performed thus: a few drops of acetic acid and 20 c.c. of a 10 per cent. solution of barium chloride are added to the specimen of urine. By this means the preformed sulphuric acid is precipitated. The solution is then filtered, thus removing the precipitate. In the filtered solution the sulphuric ether compounds will be contained together with the excess of barium chloride. On boiling the filtered solution with hydrochloric acid, the sulphuric ether compounds are split up into phenol and sulphuric acid; the sulphuric acid, uniting with the remaining barium chlorides, is precipitated as barium sulphate.

The test in itself for sulphuric acid has no diagnostic importance; but the quantitative determination of the two forms of sulphuric acid (preformed acid and sulphuric ether) is, because it gives a reliable measure of the intensity of intestinal decomposition. For this purpose the precipitate of barium (baryta) sulphate should be weighed.

Carbonates occur in solution in the urine in essential amounts only after the ingestion of fruit, etc.; they also are combined with special alkalies. Their presence is determined by the evolution of gas resulting from the addition of an acid to the urine. Concerning its diagnostic significance consult under *Reaction*, p. 183.

Ammonia appears in fresh urine, and averages in amount daily between 0.5 and 0.8 g. It is increased in many diseases of the liver and in diabetes to 6 g. Such an increase serves as a guide to determine the severity of a diabetes. The quantitative estimation is made by the addition of chloride of lime to 20 c.c. of urine, to which has also been added 20 c.c. of sulphuric acid of a definitely known composition. The mixture should be prepared in a shallow glass vessel under a bell jar. After 48 hours the change which the sulphuric acid has undergone is measured and from the amount which has been lost the formation of NH_3 is determined. Ammonia is present in large quantities in old *decomposed* (alkaline) urine; its presence may be demonstrated by holding a piece of moistened red litmus paper over a specimen of the urine, or by holding a glass rod which has been dipped in hydrochloric acid over it (cf. p. 183).

Sodium appears in daily quantities of 3 to 6 g. Na_2O; *potassium*, 2 to 3 g. K_2O. The quantitative estimation of these bodies is performed according to the rules of chemical analysis. Their presence may be of diagnostic value in some cases from the fact that in all cases of extensive albuminoid disintegration, such as occurs in fever and in inanition, the amount of potassium salts is much increased and of the sodium salts much diminished.

2. Organic Elements

Urea (designated often by clinicians with the symbol $\overset{+}{U}$) is the chief end product of the metabolism of the albuminoids.

Chemical characters. Urea crystallizes in prisms and needles, is soluble in alcohol and in water, insoluble in ether, and forms *biuret* when subjected to dry heat. This latter compound will produce a red color when added to a potassium hydrate solution containing some drops of a solution of cupric sulphate (biuret reaction, see p. 87). By the action of the bacteria in the urine urea becomes transformed into ammonium carbonate, $CO(ONH_4)_2$.

It combines with the formation of crystals, with nitric acid, and with oxalic acid.

The normal quantity of urea depends for the most part on the amount of albumin ingested (see Chap. X.); it varies between 20 and 40 g., but is less when little albuminous food has been taken and is physiologically increased after meals rich in such foods.

Pathological *increase* of the amount of urea occurs in fever, in many cases of carcinoma, in anæmia, in leucæmia, in intoxications (phosphorus, arsenic, chloroform, etc., poisoning), and in dyspnœa.

Pathological *diminution* of the amount of the urea excreted occurs in inanition, often in the diffuse diseases of the kidneys (Bright's), and in acute yellow atrophy of the liver.

Qualitative test for urea. — This test is of diagnostic importance in determining the presence of a uræmic condition, in which condition the *vomited matter*, the sputum, the effusions, and the blood may all contain urea. The fluid to be examined for this should be evaporated to the consistency of syrup; the urea is then extracted by alcohol, the solution is filtered, the alcohol is driven off by distillation, the thick liquid remaining is diluted with water, some concentrated nitric acid is added, and a portion examined under the microscope. In a little time the characteristic hexagonal crystals of urea nitrate will be observed.

Quantitative estimation of urea. — This is of great diagnostic value in cases of diseases of metabolism and in dietetics (see Chap. X.). The urea is not determined as much as is the total quantity of nitrogen which appears in the urine.

Approximate Estimation by Titration. Liebig's Method modified by Pflüger

From a graduated burette containing a solution of mercuric nitrate of definite proportions,[1] the fluid is allowed to drop gradually into a test-tube which should contain 10 c.c. of the urine. An insoluble combination of urea nitrate and mercuric oxide is formed.

---

[1] In 1000 c.c. of the solution there should be 71.48 g. of mercury. The preparation of the solution requires a great deal of care; it may be procured at many chemical factories.

Next to the burette a watch glass with a black bottom should be placed, in which a thick mixture of powdered Na_2CO_3 and water has been prepared. As long as the mercuric nitrate is united with the urea in the urine, it can not react to Na_2CO_3; but as soon as all of the urea is used up in the former combination, the $Hg(NO_3)_2$ which will now exist in a nascent condition will combine with the Na_2CO_3 to form $2\ NaNO_3$ + carbonic acid (which will arise in bubbles) + yellow HgO. The urine and the mercuric nitrate solution, as it drops into it, should be stirred thoroughly with a glass rod; the thick sodium carbonate mixture should then be stirred with the same rod; as soon as a permanent yellow color appears at the spot of the thick mixture which was touched by the rod, the titration is completed.

The number of cubic centimetres of the mercuric nitrate solution used in the test is multiplied by 0.04 to obtain the percentage of *nitrogen* in the urine. For example, 21 c.c. of the Hg solution were used in 10 c.c. of urine; $\therefore$ $21 \times 0.04 = 0.84\%$ of N. If the urine passed in 24 hours happened to be 1500 c.c., then the daily excretion of N would be 12.6 g., which equals 27 g. of urea, because the amount of nitrogen is to the urea as 1 is to 2.143 or $\overset{+}{N} : \overset{+}{U} :: 1 : 2.143$, from which $\overset{+}{U}$ was determined.

This method is for clinical purposes entirely reliable. A more exact method used requires too many details which must be carefully carried out. such as precipitating the phosphates and sulphates in the urine by a solution of baryta, the reduction of the acidity which arises in the process of titration, the repeated addition of mercuric nitrate solution, etc.

The method now used in all clinical laboratories, one which is entirely reliable, for determining the amount of nitrogen is that first recommended by *Kjeldahl*. It is carried out in this way: 5 c.c. of urine and 20 c.c. of fuming sulphuric acid are placed in a flask, which is then held over a flame until it boils, when the fluid loses its color. The solution is then diluted with 100 c.c. of sodium hydrate whose specific gravity is 1.3 and distilled in a distilling apparatus; 50 or 100 c.c. of a $\frac{1}{10}$ normal acid (sulphuric) is titrated after the distillation with a $\frac{1}{10}$ potassium hydrate solution. (All of the N of the urine is converted into NH_3, which becomes united with the excess of sulphuric acid into $(NH_4)_2SO_4$; the NaHO sets the NH_3 free, and the latter immediately combines in part with the prepared $\frac{1}{10}$ solution of normal sulphuric acid; the part which remains is determined by titration and from it is reckoned the entire amount of NH_3, from which in turn the amount of N contained in the original 5 c.c. of urine is determined).

Example: 5 c.c. of urine were oxidized; 50 c.c. of $\frac{1}{10}H_2SO_4$ were prepared; after the distillation 12 c.c. of $\frac{1}{10}$ NaOH in the titration of the acid remaining free, therefore only 12 c.c. of the $\frac{1}{10}H_2SO_4$ remained free; therefore 38 c.c. of the $\frac{1}{10}H_2SO_4$ must have been used in combining with the NH_3 which distilled over. For the combination of 38 c.c. of the $\frac{1}{10}$ normal acid, that is, 38×0.049 g. H_2SO_4, 38×0.0017 NH_3 are necessary; and the amount of nitrogen corresponding would be 38×0.0014 g., which would be equal to 0.0532 g. of N. Now since this amount was contained in 5 c.c. of urine, the entire percentage in the amount would be 1.064.

Uric acid ($C_5H_4N_4O_3$) is excreted in the urine, the daily amount being from 0.4 to 1.4 g. The quantity varies in different persons and at different times. The uric acid (together with the xanthin bases) arises from the disintegration of the nuclein which is contained in all cell bodies. The uric acid appears in the urine in solution as neutral sodium urate; in very acid and in very concentrated urine (such as occurs in fever and in congestion of the organs, in cases where a small amount of fluids is taken, and in profuse perspiration) acid sodium urate or pure uric acid is sometimes precipitated when the urine becomes cold (see urinary sediments).

Uric acid excretion is increased in all conditions of increased destruction of leucocytes, in leucæmia (the proportion in the urine is 1 to 16), after the ingestion of food rich in nuclein. In gout the excretion of uric acid varies remarkably.

The qualitative détermination of uric acid is sometimes of value in the examination of sediments and concretions (see below), and is performed by means of the *murexide* test, which is the following: —

To the specimen to be examined 3 or 4 drops of concentrated nitric acid are added in a porcelain dish and the contents of the dish evaporated to dryness; if uric acid be present in the specimen, an orange-yellow color will be produced, which will be changed by the addition of ammonia to be a purple-red.

The qualitative examination for its presence in the *blood* is carried out as was suggested by *Garrod* with a thread, constituting *Garrod's thread test.* The method is as follows: 10 c.c. of blood are obtained by wet cupping; the blood is permitted to coagulate and to settle in the cup, the serum is transferred into a watch glass, rendered acid with some 30 per cent. acetic acid, and in the mixture a thin linen thread is placed. The glass should remain covered for 24 hours, when it is examined under a low power of the microscope; should uric acid have been present, many crystals of it will be found adhering to the thread. Its presence occurs in attacks of gout and in nephritis.

To estimate uric acid quantitatively, the urine should be mixed with magnesia and silver nitrate; the resulting urates of magnesia and silver are then combined with sulphuretted hydrogen and the amount of uric acid determined by weighing. For an exact description consult the various text-books on chemical analysis.

Oxalic acid ($COOH . COOH$) is excreted to the amount of 0.02 g. in the 24 hours, either in solution or as a sediment of calcium oxalate (see below).

Xanthin bases (xanthin $C_5H_4N_4O_2$, hypoxanthin $C_5H_4N_4O$) appear in small quantities in the urine and result, like uric acid, from the disintegration of nuclein substances. The sum total of uric acid and xanthin bases are designated as *alloxan bodies*. Consult text-books for the quantitative determination. The alloxan bodies are increased in all conditions which involve augmented disintegration of leucocytes.

Kreatinin ($C_4H_7N_3O$), daily quantity from 0.6 to 1.3 g., is without diagnostic importance, as is

Hippuric acid ($C_9H_9NO_3$), whose daily amount of excretion is between 0.25 and 0.5 g., and which is formed from benzoic acid (C_6H_5COOH) and glycocol (CH_2NH_2COOH).

Indican ($C_8H_6NKSO_4$), potassium indoxyl sulphate, present in small quantities in the urine, is increased in cases of extensive decomposition in the intestinal tract, therefore in all diseases of the abdominal cavity which tend to diminish peristalsis and hinder absorption, especially in *peritonitis* and *intestinal obstruction;* at the same time the amount of indican is greater the higher the obstruction is situated. Obstruction of the large intestine gives rise to but little indican. Indican is also increased in cases of putrid suppuration.

Chemical attributes of indican. From the decomposition of albuminoids in the intestine or in foci of suppuration *indol* (C_8H_7N) is produced; indol is oxidized in the organism into *indoxyl;* like all aromatic compounds, indoxyl combines with sulphuric acid. The test for indican is based on the development of indigo blue.

Test for indican in the urine. The specimen of urine is mixed with an equal volume of hydrochloric acid; fresh chloride of lime solution is then added drop by drop while the mixture is being stirred. The chloride of lime solution should be composed of 5 parts of chlorinated lime and 1000 parts of distilled water. If indican be abundantly present, the urine will turn to a blue color, or a precipitate of indigo in the shape of blue flakes will take place. By the addition of ether or chloroform the indigo blue will be removed. Very dark urine should, before the test is applied, be mixed with some plumbic acetate and then filtered.

Indigo red is demonstrated by boiling the urine and then adding some nitric acid drop by drop, while carefully continuing the boiling process until a deep-red color appears. Should the mixture then be shaken, the froth will assume a blue-violet color, which is at once removed by the addition of chloroform or ether. This constitutes *Rosenbach's reaction.* The composition of indigo red is not as yet known; this reaction occurs quite in parallelism with the indigo blue reaction in the urine of cases of severe intestinal affections, although one or the other may be demonstrated in mild cases.

Phenols: C_6H_5OH is carbolic acid or phenol; $C_6H_4CH_3OH$ is cresol; they occur in the urine combined with sulphuric acid (sulphuric ether). In the normal state 0.017 to 0.05 g. of phenols are excreted; increase in this amount to 0.06 g. occurs in cases of decomposition occurring within the organism, in which cases the excretion of phenols offers a diagnostic measure of great value in determining the intensity of the process.

Test for phenol: 200 c.c. of urine are mixed with 40 c.c. of hydrochloric acid and distilled until 150 c.c. have gone over into the receiver; an aqueous solution of bromine is added to the distilled portion until it assumes a yellow color. A precipitate will be formed if phenol be present. The precipitate is tribromphenol. By weighing this the amount of phenol in the entire urine may be determined.

Examination of the Urinary Sediments

If the urine be very cloudy, or if it contain a sediment, pour it into a conical glass and permit the deposit to collect for several hours; then pour off the superambient fluid and place a specimen of the sediment upon the slide for microscopical examination.

For the speedy collection of a scanty sediment use a *sedimentator (centrifuge apparatus)*.

Before using the microscope, *test* the reaction of the urine and *heat* a sample so as to gain some knowledge as to its constituents. For if the urine be acid and cloudy and the cloudiness disappears on heating, the sediment will be recognized as composed of urates chiefly. Should the urine be acid and the cloudiness does not disappear on heating, but on the addition of potassium hydrate, then the sediment is due to uric acid. If the urine was alkaline and cloudy and the cloudiness disappeared on the addition of hydrochloric acid with or without the evolution of gas, then it may be considered that the sediment was composed of lime carbonate or phosphate.

Unorganized Sediments

In Acid Urine

Acid sodium urates (Fig. 35) are amorphous granules collected in clusters, usually colored yellow. They form the *brick-dust deposit (sedimentum lateritium)*, are soluble on heating as well as on the addition of potassium hydrate. On adding some hydrochloric acid to a specimen under the microscope, crystallization of uric acid may be observed. They are of no diagnostic importance, and indicate only the acid condition and concentration of the urine.

Uric acid (Fig. 35) occurs in the shape of whetstones, and in cylinders (also in the shape of spears and rosettes) chiefly of a yellow color; it dissolves on the addition of potassium hydrate, but not on heating. It is recognized outside of the shape of its crystals by the murexide reaction (see p. 207). An abundant deposit of uric acid does not in itself denote that the uric acid is increased, but often only that the conditions of solubility are unfavorable, either because there is too small an amount of water, or because there is too much acid in the urine; still, when abundantly present, it should demand a quantitative analysis. An abundant uric acid deposit in the urine indicates the presence of the so-called condition of uric acid diathesis (*nephritis urica, arthritis urica*).

Fig. 35. — Urinary Deposits in Acid Urine.

Oxalate of lime (Fig. 35) (the reaction of the urine is nearly neutral) forms in octahedral crystals. If present only as solitary crystals, it has no significance; even an abundance of them in the urinary deposit does not always indicate an increase in the excretion of oxalic acid; to determine this a quantitative analysis is necessary.

Cystin occurs rarely as a deposit. When it does it is pathognomonic of a special form of metabolic disturbance (see Chap. X.). It consists of hexagonal crystals which are readily soluble in ammonia.

Leucin (amido-capronic acid) (Fig. 36) and *tyrosin* (amido-paracumaric acid) are likewise rarely found as urinary deposits; they occur in the urine in cases of acute yellow atrophy of the liver and in poisoning by phosphorus. Leucin crystallizes in yellow-white crystals which are often in the shape of striped radiatory globules; tyrosin crystallizes as beautiful needle-like bundles.

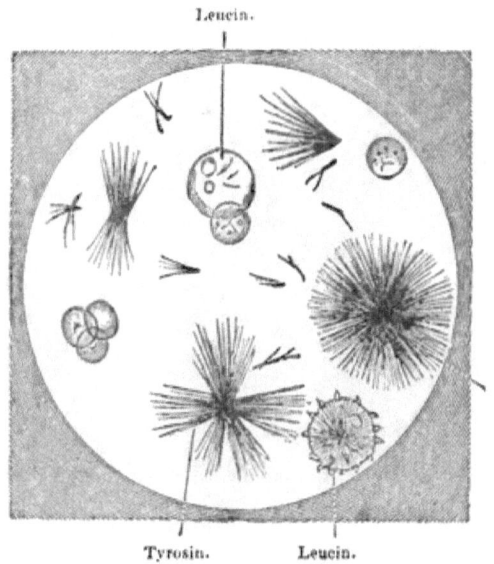

FIG. 36.—SEDIMENT OF URINE IN ACUTE YELLOW ATROPHY OF THE LIVER.

In Alkaline Urine (Fig. 37)

The *ammonio-magnesia phosphates* or *triple phosphates* ($NH_4MgPO_4 + 6 H_2O$) crystallize in coffin-like crystals which are very readily soluble in acetic acid.

Phosphate of lime occurs either as $Ca_3(PO_4)_2$ in the form of irregular granules, or as $CaHPO_4$ in cuneiform crystals, which are usually collected together in masses and often in rosettes.

Carbonate of lime ($CaCO_3$) occurs as round regular granules or in the shape of dumb-bells, and dissolves on the addition of an acid with an evolution of gas.

Ammonium urate occurs in the shape of pine cones, or in irregular fusiform crystals.

The alkaline deposits, outside of indicating the reaction of the urine (see above), have no farther diagnostic significance.

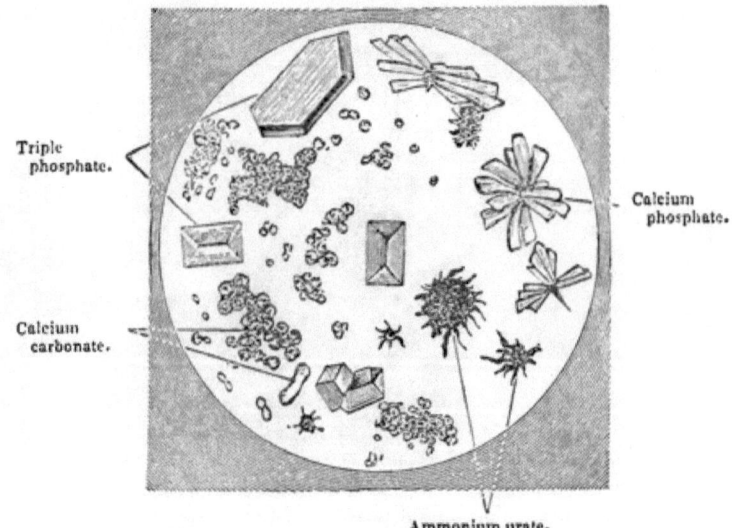

Fig. 37.—Sediment of Ammoniacal Urine.

Organized Sediments (Figs. 38 and 39)

These are of the utmost importance in the diagnosis of diseases of the kidneys (see Chap. IX.).

White blood corpuscles (leucocytes) appear sparsely in the urine of healthy persons; should they be abundant, however, it indicates an inflammation or a suppuration in some spot along the genito-urinary tract, a nephritis, pyelitis, cystitis, gonorrhœa, or a leucorrhœa in women.

Red blood corpuscles occur mostly as pale corpuscles, owing to imbibition of water; they indicate the presence of a hæmorrhage somewhere in the genito-urinary apparatus (see hæmaturia, p. 189).

Renal epithelium (Fig. 38) occur as round or cubical cells containing nuclei, and indicate chiefly an affection of the kidney. They often coalesce into *epithelial casts*. *Fatty* epithelial cells (masses of fat granules) are of the utmost diagnostic importance (Fig. 39); they indicate chronic parenchymatous nephritis in the 2d stage of a fatty degeneration.

Epithelium from the renal pelvis, of the *ureters*, and of the *bladder* can not be distinguished from each other; they occur either as flat polygonal, or more or less round cells, provided with processes and in part containing nuclei. An abundance of these cells in the urine justifies the diagnosis of a pyelitis, of a cystitis, or of an inflammation of the ureter (see differential diagnosis, Chap. IX.).

Epithelium from the vagina consists of large flat epithelium like the buccal epithelium; *epithelium from the male urethra* is of the cylindrical variety, and appears sometimes in gonorrhœal pus.

Casts in the urine are probably exudations from the renal tubules. They appear as:—

1. *Hyaline casts,* which are small, transparent, homogeneous formations, with not very definite contours. Their presence is not a proof of nephritis. They occur also in congestion, in fever, and in jaundice, and even in very small quantities in the healthy person.
2. *Epithelial casts,* consisting of agglutinated epithelial cells, indicate the presence of a nephritis; they are often deformed, more or less granular (*granular casts*), and often are covered by fatty epithelial cells.
3. *Blood casts* occur only in hæmorrhage of the kidneys.
4. *Waxy casts* have sharp outlines, a yellow glistening appearance, and occur only in chronic nephritis.
5. *Brown casts* occur rarely in severe infectious diseases, and in fractures of bones.

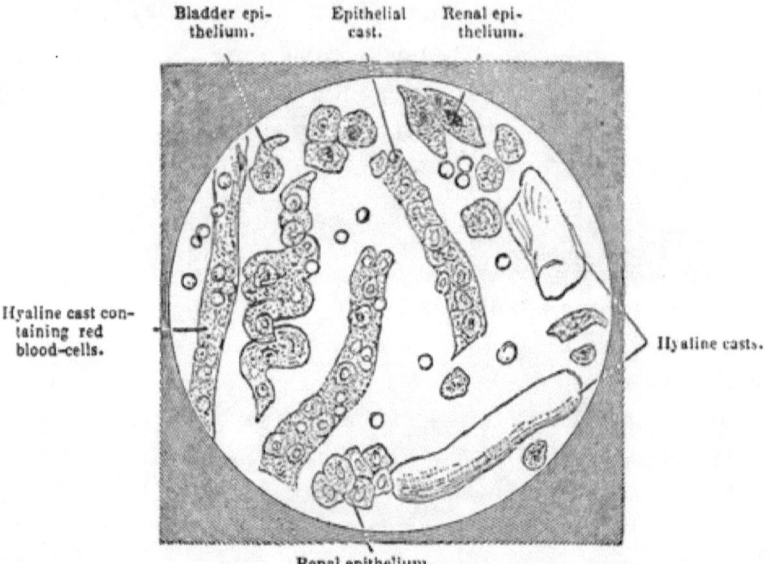

Fig. 38.—Sediment in Acute Nephritis.

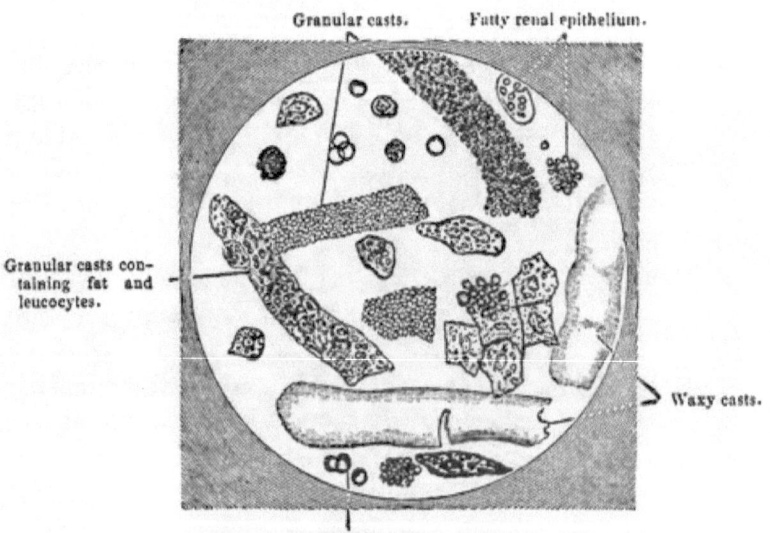

Fig. 39.—Sediment in Chronic Nephritis.

Cylindrical formations which look like casts are often seen; they are produced by accumulations of bacteria or of amorphous urates.

Micro-organisms may occur in the urine in many infectious diseases (diphtheria, relapsing fever, and in typhoid fever). The presence of gonococci and of *tubercle* bacilli have of course a diagnostic significance. In cystitis and in pyelonephritis an abundance of microbes appear in the freshly voided urine. (For their demonstration, see Chap. XII.)

Parasites have been observed in rare cases in the urine in the shape of particles of *echinococcus*, embryos of *Filaria sanguinis*, which, like the *Distoma hæmatobium*, have given rise to hæmaturia.

Tests for Certain Heterogeneous Substances occurring in the Urine

The determination of the presence of foreign substances in the urine is of importance in diagnosing the various *intoxications*. Irrespective of this, it is often of interest from a therapeutic standpoint whether a substance has been absorbed by the organism. Finally, by demonstrating the presence of a medicine in the urine, we may be able to form opinions concerning the patient's statements.

Iodine appears in the urine after the use of potassium iodide and iodoform. Add to the urine a few drops of fuming nitric acid, or of an aqueous solution of chlorine, and a few cubic centimetres of chloroform; then shake the test-tube containing the mixture; should iodine be present, the chloroform will assume a red-violet color.

Bromine appears in the urine after the use of the bromides. Test in the same manner as for iodine. When large quantities are present, the chloroform will become yellow. The test will not detect a small quantity. In the latter case, we render the urine alkaline with sodium carbonate, and add to it 2 g. of potassium nitrate. Pour the mixture then into a platinum dish, and evaporate and heat the dry residue until it melts. The residue, when cooled, is dissolved in water rendered strongly acid with hydrochloric acid, and is then shaken with chloroform. With this test, if the urine contains only a very small quantity of bromine, the chloroform will become yellow.

Iron. — Urine containing much iron will become greenish-black on the addition of ammonium sulphide. To test the presence of only a small quantity of iron, 50 c.c. of urine are evaporated in a platinum dish, the dry residue is converted into an ash, and then the ash is subjected to the action of a weak solution of hydrochloric acid. Should iron be present, a blue precipitate will result after the addition of some potassium ferrocyanide.

Arsenic. — In order to successfully carry out the test for arsenic, the organic substances in the urine must be removed. This may be done as follows: one or two litres of urine are evaporated in a porcelain dish to $\frac{1}{8}$ of its volume; then an equal volume of concentrated hydrochloric acid is added to the residue and the mixture covered and placed on a water bath. At regular intervals during the heating on the bath, 2 or 3 g. potassium chlorate are added, until the mixture becomes entirely light yellow. It is then rapidly heated and boiled until all smell of chlorine disappears, when it is strongly diluted with water. For many hours after this, sulphuretted hydrogen is passed through the solution, producing a precipitate of arsenic sulphide, which is removed by filtration, dried, dissolved in a small dish by a mixture of a few drops of nitric and sulphuric acids, heated until the acid smell has disappeared, then strongly diluted with water, and the remaining solution is then subjected to the *Marsh mirror test* for arsenic.

The *Marsh test* is performed in the simplest way, as follows: a test-tube is provided with a cork in which a hole has been bored; into the hole one end of a glass tube — which has been bent to a desirable angle, and whose other extremity has been drawn out to a capillary point — is introduced. The test-tube is now filled with zinc and dilute sulphuric acid, and with some of the previously prepared solution of urine. Gas will be immediately evolved, when the cork and attached glass tube should be introduced into the mouth of the test-tube. When the gas is streaming off strongly, the gas at the capillary end of the tube should be ignited by a match. If arsenic be present in the solution, the gas will burn with a blue flame, and if a clean, dry, porcelain plate be held over the flame, a mirror of deposited arsenic will form on the plate. This mirror has a metallic, glistening appearance, and will be dissolved by a solution of sodium hypochlorite.

Lead. — The organic substances should be disintegrated by hydrochloric acid and potassium chlorate, as was done in the test for arsenic (see above), and, through the diluted mixture, which has been thereby rendered weakly acid, sulphuretted hydrogen is permitted to pass; this will form a precipitate of the brownish-black plumbic sulphide when lead is present.

Mercury. — About 1 litre of urine is heated to 60° to 80° C., made acid by the addition of hydrochloric acid, and then thoroughly oxidized by the addition of fine copper filings (lametta). After an hour, the urine is decanted; the copper filings are washed with hot water, then with alcohol, and finally with ether, and thoroughly dried between layers of filter paper. The filings are then placed in a small glass tube, which is thereafter drawn out at both ends into capillary points. The tube is then heated in the flame, the mercury becomes sublimed, and in the capillary ends metallic, silver-like rings will have formed. If a small piece of iodine has been put in the tube before the ends of the tube have been drawn out, the red mercuric iodide would be formed instead.

Carbolic acid. — If much carbolic has been taken, the urine will have a greenish-brown appearance, and by exposure to the air it will become darker. The same color results from the ingestion of *hydroquinon* and from *uva ursi*. The test for carbolic acid is made by adding bromine water to the distilled mixture (see p. 209), or by the test for the presence of the sulphuric ethers (see p. 203).

Quinine. — Shake well a large quantity of urine, to which some ammonia has been added, with ether. This will take up the quinine. The residue left after evaporating the ether is mixed with acidulated water. The mixture is then subjected to the action of chlorine water, and after that to ammonia. A green color will develop if quinine be present.

Salicylic acid ($C_6H_4OH \cdot COOH$). — The urine will give a blue-violet color if ferric chloride be added. Should the reaction to this be negative, acidulate in a graduate 30 c.c. of urine with sulphuric acid, and add to it 30 c.c. of ether; then shake the mixture thoroughly, pour off the ether, and to the latter add some of a ferric chloride solution drop by drop. A blue color will ensue even where the quantity of salicylic acid present is very small.

Antipyrin. — A red color is produced by the addition of ferric chloride to the urine.

Antifebrin. — A red color is produced when the urine is boiled with hydrochloric acid and afterwards mixed with some of a 3 per cent. carbolic solution and with ferric chloride; on the addition of ammonia, it changes to a blue color.

Phenacetin. — A brownish-red color will ensue on the addition of ferric chloride.

Tannin. — With ferric chloride the urine becomes dark blue, almost black.

Napthalin. — After the ingestion of large doses, the urine will be turned to a green color by the addition of concentrated sulphuric acid.

Turpentine. — The urine has an odor of violets.

Rhubarb and **senna** (chrysophanic acid). — The urine becomes purple-red on the addition of sodium hydrate. Sodium carbonate will also produce the same reaction.

Santonin. — The urine, when santonin has been taken, has a straw color, which is turned to red by the addition of sodium hydrate. Sodium carbonate, however, will *not* produce the red color.

CHAPTER IX

DIAGNOSIS OF THE DISEASES OF THE KIDNEYS

The Diffuse Diseases of the Kidneys (Bright's Disease)

The diffuse diseases of the kidneys are recognized by the simultaneous presence of *œdema* and *albuminuria;* they are classified under the head of *Bright's disease*. The specific diagnosis of the form of Bright's disease present is decided chiefly by the examination of the urine (see Chap. VIII.), and, in addition to this, by the history, the course of the disease, and by the examination of the other organs (heart, vessels, liver, spleen, and the eyes).

In reference to the *history*, the following data are especially important: alcoholism often leads to the production of a chronic nephritis; chronic intoxications and gout lead frequently to the primary atrophy of the kidney; severe exposure, frequent drenching of the skin, the action of the toxic agents, especially the acute infectious diseases (scarlatina, etc.), are potent in producing acute nephritis. *Tertiary syphilis, suppuration, phthisis,* and *malaria may lead to amyloid degeneration of the kidney.* Inquiry should always be made as to the duration and course of the disease, and the investigation should elicit whether there have been previously any nephritic symptoms, such as œdema, urinary changes, headaches, vomiting, asthma, and visual disturbances.

Concerning œdema, see p. 10. For the tests for albumin, see p. 186.

In many periods and in many forms of Bright's disease either œdema or albuminuria may be absent, and not alone this, but in some rare cases both may for a time be wanting. Under these circumstances the consideration of the course of the disease, the examination of the heart (cardiac hypertrophy in atropic kidney) and of the pulse, and often the large quantities of urine will aid the diagnosis. In spite of these exceptions it is wise to establish the rule that œdema and albuminuria are the cardinal symptoms of diffuse diseases of the kidneys.

In order to make a diagnosis of the form of Bright's disease present in each case, it is best to fix in the mind a systematized scheme of the various forms of the disease which occur.

BRIGHT'S DISEASE
ŒDEMA WITH ALBUMINURIA

	INFLAMMATORY FORM	NON-INFLAMMATORY FORM
First stage ..	Acute hæmorrhagic nephritis.	Œdema from venous stasis (chronic congested kidney).
Second stage .	Chronic nephritis; fatty degeneration.	Amyloid degeneration.
Third stage ..	Secondary atrophy; white atrophic kidney.	Primary atrophy (red atrophic kidney; arterio-sclerotic, lead, and gout kidneys).

In a class by themselves are the œdema and albuminuria of pregnancy — the kidneys of pregnancy.

In the first column of this table or scheme the forms of nephritis are mentioned, as well as how they sometimes develop in order from one form to the other; still the chronic form may occur *primarily* without there having been an acute inflammation before its development.

Clinical diagnosis must renounce deciding between the presence of an *interstitial* and a *parenchymatous* inflammation, because the symptoms do not in most respects harmonize with the pathological anatomical processes. Still the proof of the existence of fat in the urinary deposit is a positive indication of a parenchymatous process, while, on the other hand, the absence of an essential sediment in the urine would point to an interstitial inflammation.

In the second column those forms are mentioned which exist individually and which are *not related to each other*, but which in their clinical features very much resemble the forms adjoining them in the first column.

The chronic congested kidney strictly should not belong to this group, but clinically it gives a symptom picture so closely resembling the essential forms of Bright's disease that it appeared to the author justifiable to place it in the table.

Chief Symptoms of the Various Forms of Bright's Disease

Acute hæmorrhagic nephritis presents the following clinical features: generally severe anasarca, and especially œdema of the face. The urine contains much albumin and a good deal of blood. Its quantity is much diminished, and its specific gravity is high. The sediment will be found to consist of red blood corpuscles, hyaline and granular casts as well as blood and epithelial cells.

The diagnosis should include an inquiry into the etiological factor: is it a nephritis due to an infection, to a poison, or to exposure? or is it an exacerbation of the chronic form? The prognosis is essentially influenced by the quantity of urine and the presence of uræmic symptoms (headache, vomiting, coma, and convulsions).

Chronic nephritis, fatty degeneration. — Usually extensive œdema, accompanied by a large amount of albumin. The quantity of urine varies. The urinary sediment is characteristic and contains *globules of fatty granules*, granular and often waxy casts, and much epithelium. The disease is usually fatal in from one to two years, in an attack of uræmia or through complications (rupture of the œdematous skin and erysipelas, pneumonia, etc.).

Primary atrophic kidney. — Abundant urine of low specific gravity, greenish-yellow of color, with little albumin, very little or no sediment. *Hypertrophy of the heart*, pulse of very high tension, retinitis frequent; begins insidiously and often develops very slowly. It occurs in general arteriosclerosis (arterio-capillary fibrosis), in gout, and in plumbism (uric acid kidney).

The atrophic kidney may also develop from a previous nephritis of long standing or from a puerperal nephritis. Rarely, it may arise from chronic congested kidney. The development of a secondary atrophic kidney from a chronic parenchymatous nephritis is exceedingly rare.

Chronic congestion of the kidney presents œdema, chiefly in the legs, cyanosis, and dyspnœa. There is a coexisting cardiac or pulmonary affection. The urine, scanty and dark-colored, is of a high specific gravity. It shows the brick-dust deposit (*sedimentum lateritium*), and contains but little albumin.

Amyloid degeneration of the kidney proceeds generally with the symptoms of a chronic nephritis. The proof of the etiological factor is necessary to a diagnosis (see "history"), as is also the simultaneous enlargement of the liver and spleen.

The other diseases of the kidney are not accompanied with œdema, but often with changes occurring in the urine. They are recognized partly by these features and partly by the occurrence of pains in the region of the kidneys, by palpation and percussion of the abdomen and especially of the renal region.

The following are chiefly to be considered: hæmorrhagic infarction of the liver, pyelonephritis, renal calculi, renal tuberculosis, tumors of the kidney, and movable kidney.

Pains in the region of the kidneys (in the loins) appear so frequently in entirely different diseases that this symptom should be used only with reserve in making a diagnosis of kidney disease. Paroxysmal severe pain (*renal colic*) is the sign of the presence of renal calculi.

Situation of the Kidneys, Percussion

The kidneys lie between the 12th dorsal and 3d lumbar vertebræ. The right kidney reaches to the liver above, the left to the spleen.

The object of percussion of the renal region is to locate the lower and outer borders of the kidneys; however, the results of this are not entirely reliable, on account of the large fat accumulation which sometimes envelops the capsule of the kidneys and on account of the varying condition of the intestines as to solid contents.

Marked increase of renal dulness occurs when the kidney is enlarged by the development of a tumor of the organ (see below); the entire absence of renal dulness justifies the conclusion that a movable kidney (floating kidney) is present, which is found much more frequently on the right side than on the left.

Chief Symptoms of a Few of the Non-diffuse Diseases of the Kidneys

Hæmorrhagic infarct of the kidney is characterized by the sudden appearance of *hæmaturia;* pain in the region of the kidney and a slight rise in temperature. The diagnosis is aided by the proof of the etiological factor, an embolism due to a valvular defect or cardiac dilatation. The hæmaturia disappears quickly.

Suppurative nephritis (pyelonephritis). — There is an irregular remittent type of fever associated with chills; the urinary sediment contains pure *pus*, usually without casts; there is blood in the urine only when the pyelonephritis has been caused by trauma or renal calculus. Severe pain in the renal region is frequent.

Tumor of the kidney is determined by palpation, bimanual palpation being often necessary; it is often very difficult to determine the renal origin of the tumor, the diagnosis being aided by the immobility of the tumor on forced respiration, the constriction of the colon, in certain cases by the superposition of the distended large intestine over the tumor, and the increased area of renal dulness.

We must decide differentially between an echinococcus (positively determined when the hooklets or the cyst wall is found in the fluid aspirated from the tumor), hydronephrosis (determined by the intermittent filling and emptying of the fluctuating sac and by the presence of urea in the fluid gained by aspiration), and carcinoma and sarcoma (they are solid tumors and are accompanied by a rapidly progressing cachexia).

Renal calculi (nephrolithiasis) is characterized by attacks of renal colic which terminate with the passage of the concretion; by the excretion of sediment independent of the attack of colic, the sediment being chiefly composed of phosphates or urates; by the frequent occurrence of hæmaturia. The reaction of the urine varies according to the nature of the calculi.

Movable kidney. — On palpation a movable abdominal tumor of the shape of the kidney is felt. The area of kidney dulness is obliterated. There are numerous nervous complaints, the patient complaining especially of traction and weight in the abdomen.

Bladder diseases present for diagnosis these conspicuous forms: —

Cystitis, which is diagnosed by the presence of frequent, painful vesical tenesmus, associated with the voiding of dirty cloudy urine full of pus cells and frequently the seat of ammoniacal decomposition (see pp. 185 and 212). The cause of the cystitis must be determined (gonorrhœa, urethral stricture, hypertrophy of the prostate, stone in the bladder, *paralysis of the bladder from spinal cord disease*). In acute cystitis (the result of exposure or of gonorrhœa) the urine is scanty, cloudy, acid in reaction, bloody, and there is intense local pain and spasm.

Stone in the bladder is diagnosed by the presence of frequent hæmaturia without a characteristic sediment, disturbances in passing urine, in many cases an accompanying cystitis and pains which radiate to the end of the penis. The diagnosis is only positively made by sounding the bladder.

Tumors of the bladder (papilloma [papillomatous fibroma, villous disease] and carcinoma) are diagnosed by the occurrence of attacks of pain and hæmorrhages in the bladder, and by the presence of a chronic cystitis. The diagnosis is only established where a tumor can be felt by a digital examination in the rectum or vagina, or by the catheter introduced into the bladder, or where particles of the tumor can be demonstrated to exist in the urine. Finally the *endoscope* or *cystoscope* may show the presence of a growth.

The Examination of Renal and Vesical Calculi

We distinguished between: 1. *Uric acid* calculi, which are most frequent in occurrence; they are hard, their surface is smooth or only a little nodular, their color varies from a yellow to a reddish-brown. Calculi composed of *ammonium urate* are brittle, and of a dirty grayish-yellow color. 2. *Oxalate of lime* calculi (mulberry calculi) are very hard, they have a rough warty surface, and vary in color from brown to black. 3. *Phosphatic* calculi (composed of calcium phosphate and of ammonio-magnesium phosphates) are soft and easily pulverized; their surface is sandy and rough, often glistening; their color is usually white. 4. Calculi formed of the *carbonates* are hard, like chalk, with a smooth surface and of a white color. 5. *Cystin* calculi are as a rule small, tolerably hard, smooth, and yellowish. 6. *Xanthin* calculi are quite hard, of a cinnamon-brown color, and their surface when rubbed glistens like wax.

Frequently the concretion or stone is not composed of one material alone, but the nucleus may be composed of one chemical compound and the body of the calculus of another.

Their composition is determined by a chemical analysis like the following, which is recommended by *Salkowski*.

It is advisable for conducting the analysis to determine whether the calculus is composed of organic or of inorganic materials. If it be of an organic compound, the calculus when pulverized will be completely consumed when heated in a platinum capsule or it leaves but a very little ash residue. In this case the calculus is composed of uric acid, of ammonium urate, of cystin, or of xanthin. Should an ash residue be left after heating, the calculus may be composed of urates, phosphates, or oxalates.

I. The Uric Acid Calculus is Completely Consumed by Heat

The powdered calculus is digested with a weak solution of hydrochloric acid and gentle heat.

a. The portion of powdered calculus used will either be dissolved completely or for the greatest part: in this case the calculus is composed of cystin or xanthin. *Cystin* will be dissolved by ammonia and will crystallize after the evaporation of the solution into hexagonal crystals.

Xanthin is tested by dissolving a portion of the powdered calculus in a porcelain dish with some nitric acid and slowly evaporating the mixture. If xanthin be present, a lemon-yellow residue will be left which is *not* soluble in ammonia, but which will change to a reddish-yellow on the addition of some potassium hydrate.

b. Or the portion used will not be dissolved completely, in which case the solution is filtered, and the filtrate is tested for *uric acid* by the murexide test: the filtrate may contain ammonium chloride, which may be tested by adding some sodium bicarbonate and heating the mixture, when the odor of *ammonia* will be detected; or some moistened red litmus paper may be held over the dish containing the mixture and the red litmus paper will become blue; or a glass rod dipped into hydrochloric acid may be held over the dish and a cloud will form above the dish about the rod.

II. The Calculus may turn Black, but may not be Consumed

Another portion of the pulverized stone is digested with some weak hydrochloric acid and exposed to heat. Should the powdered stone effervesce, it contains a *carbonate*.

a. Complete solution of the contents of the dish denotes an absence of uric acid.

b. Incomplete solution denotes that the residue may contain uric acid or an albuminoid substance. This may be determined by the murexide test.

The filtered solution is rendered slightly alkaline by the addition of ammonia; it is then rendered slightly acid again with acetic acid. Should a white powdery precipitate form, it denotes the presence of *oxalate of lime*. This is removed by filtration, and the filtered solution should be tested for phosphoric acid, for calcium, and magnesium. A portion of the filtered solution is mixed with some ferric chloride, which, if *phosphoric acid* be present, will produce a grayish-white precipitate. The chief remnant of the filtered solution should then be mixed with ammonium oxalate, whereupon a precipitate being formed would denote the presence of *magnesia*. The lime is removed by filtration and to the filtered solution some of a solution of sodium phosphate is added and the whole subjected to heat. It is then rendered alkaline with ammonia. A slowly developing crystallizing deposit will indicate the presence of *magnesia*.

CHAPTER X

THE DIAGNOSIS OF THE DISORDERS OF METABOLISM

Laws of Normal Metabolic Processes

The human body requires the ingestion of food material in order that, on the one hand, it may conduct its vital functions undisturbed, and on the other hand, that it may not consume the essential constituents of its own tissues, which consist of the albuminoids, fats, carbohydrates, water, and inorganic salts.

The products of the disintegration of the albuminoids are excreted by the urine as urea, uric acid, etc.; the fats and carbohydrates are oxidized into carbonic acid and removed from the body in the breath, but they are nevertheless of great importance in the preservation of the organism.

The albuminoids are very complex bodies, whose chemical composition has not as yet been sufficiently investigated; only this much is known, that from the disintegration of albumin occurring in the body different chemical groups are formed, viz.: (1) a nitrogenous, urea-like group, which is excreted in the urine; (2) an aromatic group containing or represented by C_6H_5, which is likewise excreted in the urine; (3) a group resembling the fats, and which further *act like fats* and *carbohydrates* and are removed from the body by the breath as CO_2. In this way the positive fact is explained that *fat and sugar can be formed in the body from the albumins*, and that albumin, when introduced in sufficient quantity, can replace the other nutritive materials. On the other hand, the fats and carbohydrates, which lack the nitrogenous and aromatic compounds, can only replace the albumin to a limited extent.

The requisite amount of food-stuffs. — In order that the nutrition of the body may be kept at the normal, it requires that the body be given a certain *amount* of food. In order that in estimating the amount of nourishment needed, a uniform *measure* may be used for the different food-stuffs, we make use of the *amount of heat* which is developed from the disintegration of food, and which has been used by different investigators in their experiments and investigations. Therefore as the unit value we make use of the *calorie*, which is that amount or quantity of heat which is necessary to raise one kilogramme of water one degree in temperature.

We may therefore substitute for the amount of food-stuff its corresponding value in calories, in this manner: —

   1 g. albumin = 4.1 calories
   1 g. fat   = 9.3 calories
   1 g. carbohydrate = 4.1 calories
   1 g. alcohol = 7.0 calories

Instead of saying a strong man needs for nourishment 118 g. of albumin, 56 g. of fat, and 500 g. of carbohydrates, we may say he needs the equivalent of 3054.6 calories.

The number of calories which are needed for nourishment by a healthy man depends on his body weight, on the amount of work performed by him, and on his previous condition of nutrition. A healthy, strong laborer weighing about 70 kg. requires daily about 3000 calories, and when doing work requiring the exercise of additional effort, the necessary number of calories will rise to between 4000 and 5000; a less strong laborer, whose weight is about 50 kg., requires about 2400 calories. The requisite amount therefore for a healthy man is about 45 calories for every kilogramme of weight. However, it is entirely improper, if we wish to determine the amount of nourishment needed by a healthy man, to simply multiply his body weight by a certain number of calories. Moreover, the necessary number of calories necessary for his sustenance depends essentially on the condition of the physical changes of the days immediately preceding. For instance, if a man, otherwise normal, have an œsophageal stricture of long standing which has left him badly nourished, and by reason of this much emaciated, he may retain the condition in which he is at the time, even with from 1000 to 1500 calories, yes, even with less. To estimate the number of calories necessary, it is essential to study in each individual case the nutrition of the body and conditions of the preceding days.

Relations of the food-stuffs to each other. — It is very essential to observe in nutrition that the representation of the various food-stuffs corresponding to their caloric values is *only possible up to certain limits*. It is, moreover, necessary that a definite amount of *albumin* be always administered to the body; *this can not be replaced by fats and carbohydrates*. The quantity of this necessary amount of albumin (*albumin necessary to nutrition*) depends upon the condition of nutrition, rather on the store of albumin in the body, and, on the other hand, upon the amount of carbohydrates and fats which have at the same time been administered.

The amount of albumin necessary for the nutrition of a strong, well-nourished man is from 80 to 100 g.; in the poorly nourished and in the non-worker it may be less.

It is only when the amount of albumin necessary for the body's nutrition has been reached, that the food-stuffs may be reckoned entirely according to their corresponding caloric values; and it depends more upon the condition of the stomach and the digestion whether fats and carbohydrates or whether more albumin should be given.

Before the power of substitution of one food for another, by virtue of the measurement of its calories, was sharply defined, it was known that the various foods could replace each other under certain conditions; these were called *isodynams*. Thus 100 g. of fat equalled 211 g. of albumin, or 232 g. of starch, or 234 g. of cane-sugar, or 256 g. of grape-sugar.

Exchange of food-stuffs. — The *transposal of the albumin* depends upon the amount of nourishment taken, and as well, indeed, upon the number of its calories and upon the amount of albumin taken in the food. If less albumin is administered, *where there is a sufficient number of calories*, than is contained in the albumin needed for nutrition, then more nitrogen is excreted in the urine than is contained in the food (1 g. of N corresponds to 6.25 g. of albumin). Should albumin be contained in the food given to cases where the total number of calories is sufficient, then an *equilibrium* is *formed* in the *nitrogenous production*, that is, the excreted N is equal to the amount of N administered. Should more albumin than is necessary be administered, in cases where the total number of calories is sufficient, then more N is excreted, and the equilibrium as far as N is concerned is soon reëstablished.

Considerable proteid increase can be attained only in a person in the process of growth, in cases of convalescence from an acute disease, and in simple inanition.

If the total number of calories contained in the food is not sufficient, an increase of the excretion of nitrogen may take place even in cases where sufficient albumin has been taken as food.

Again, the albuminous exchange depends upon the previous nutrition and the condition of the body produced by that nutrition. Muscular people having plenty of albumin consume more albumin than fat people in whom the exchange of albumin is less.

Work has no influence on the consumption of albumin in general; but it favors the consumption of fats and carbohydrates. Should not sufficient fats and carbohydrates be administered in food, the body will then consume its own fat, in order that the necessary work may be done.

However, everything in such cases depends upon the total number of calories; if it be very large, the contribution to the amount needed for work, where there has been no fat or carbohydrates administered, is made good in part by the metabolism of the albumin itself into the fat groups; and if the number of calories be insufficient, then both the fat as well as the albumin of the body are consumed.

Anomalies of Metabolism

The anomalies of metabolism thus far known consist of:—

1. *Qualitative changes* by which the urine will contain substances which are not excreted in it in health.

The most important qualitative changes occur in *diabetes mellitus*, where grape-sugar appears in the urine, while in health all the carbohydrates are transformed in the organism into CO_2.

Certain forms of obesity seemingly depend upon the loss of the power to oxidize the fat formed in the body.

In some rare forms of disordered metabolism, which are not yet sufficiently known, some extremely peculiar substances are excreted in the urine; for example, *cystin* and *diamine* in cases of cysturia.

2. *Quantitative changes.* These are chiefly shown in the metabolism of the albuminoids. The laws of the equilibrium of N where the income is sufficient, which were mentioned in the previous paragraphs, suffer in a few diseases an alteration in the shape of a more extensive transformation: an increased consumption of albumin, a more marked oxidation of the albuminoids of the body takes place. This is noticed in fever, in many cases of phthisis, in carcinoma, in anæmia, and in leucæmia.

A diminution of proteid metabolism occurs in cases of convalescence from acute disease, in inanition, in many forms of obesity, and in myxœdema.

The conditions of diminished excretions as a result of disease of excreting organs, as, for example, the diminution of the excretion of urea in *nephritis*, belong to the disturbances of metabolism in the broader sense. The diseases of the pancreas, liver, and intestines ought to be here mentioned; in these conditions less fat and albumin are absorbed in the intestine than under normal conditions of the organs.

In gout it would appear that the question is one of an increased production of uric acid from an increased destruction of nuclein (see p. 206); the excess of uric acid in the blood can be partly accounted for by inflamed and necrotic areas of tissue (gout necrosis).

In order to diagnose a disturbance of metabolism with certainty, it is necessary to tabulate the income and the expenditure as in a balance-sheet. Clinically, the determination of the following data is sufficient: —

1. *The value of the food.*
2. *The constituents of the urine* (N, sometimes uric acid, finally sugar).
3. *The unabsorbed portion of food* in the faeces, determined by the amount of N and fat contained therein.

By determining these points the metabolism of the albuminoids can be accurately accounted for; the metabolism of the fats and carbohydrates will escape a quantitative estimation if the amount of carbonic acid in the breath be not measured.

1. The Value of the Food

In order to accurately determine this, it is necessary that everything which the patient eats should be weighed by the scales and what is left uneaten should be deducted. The value of the different articles of food in nutritive materials is cleared up by the following table: —

Articles of Food	Albumin Per Cent.	N Per Cent.	Fat Per Cent.	Carbohydrates Per Cent.	Analysis made by
Raw beef, free from visible fat	18.36	3.4	0.9	—	Voit.
Moderately fat ⎫	20.91	3.3	5.19	0.48	König.
Fat, uncooked ⎬ Beef .	17.19	2.8	26.38	—	König.
Roast ⎮	30.56	4.89	6.78	—	Rubner.
Boiled ⎭	21.8	3.5	4.52	—	Renk.
Roast ⎫ Veal	18.88	3.02	7.41	0.071	König.
Raw ⎭	15.3	2.84	5.2	—	Renk.
One egg (45 g. without shell)	6.25 g.	1 g. N	4.9 g.	—	Voit.
Good milk . . .	4.13	0.64	3.9	4.2	Voit.
Milk (for children) (Charité)	3.88	0.62	3.1	4.5	The author.
Skimmed milk . . .	3.25	0.52	1.1	4.1	The author.
Butter	0.5	0.08	87.0	0.5	König.
Cheese	32.2	4.75	26.6	2.97	Renk.
Bacon (Charité) . .	—	—	94.7	—	The author.
White bread (rolls) .	9.6	1.5	1.0	60.0	Renk.
Rye bread, fresh . .	5.63	0.9	—	44.0	The author.
Bread (Charité). . .	8.22	1.315	0.64	58.3	The author.
Boiled potatoes, without the skins . . .	2.18	0.35	—	23.0	Rubner.
Vegetables (Charité) from 3 analyses . .	3.45	0.55	4.2	20.3	The author.
Soup (Charité) from 3 analyses	1.7	0.272	1.8	8.3	The author.
Beer (light)	0.56	0.09	—	5.5	The author.
Wine	0.19	0.03	—	2.0	König.
Coffee (weak) . . .	0.25	0.04	—	—	The author.

2. The Constituents of the Urine

It is of prime importance to collect the urine passed in the 24 hours, without loss. The total amount of nitrogen contained therein is then determined according to the rules given on p. 206.

In *diabetes* a quantitative analysis of the sugar passed must be made (p. 196).

3. The Undigested Nitrogenous Materials and Fats left in the Fæces

The fæces passed on the day are marked by previously administering some of a black coloring charcoal mixture. The fæces are dried; the nitrogen in them is determined by the method of *Kjeldahl*, and the fat by extracting with ether.

It is customary to add the N in the fæces to the N in the urine and to record in the table under food the added figure as the expenditure.

The estimation of the amount in the fæces is both tedious and laborious; we can not do without it in some cases. For clinical purposes the values formed by Rubner as representing the consumption of food in the intestines may be used with advantage.

Article of Food	N Per Cent.	Fat Per Cent.	Carbohydrates Per Cent.
Meat	2.65	19.2	—
Eggs	2.9	5.0	—
Milk	8.9	5.7	—
Wheat bread	20.7	—	1.1
Rye bread	32.0	—	19.9
Potatoes	32.2	—	7.6
Vegetables	18.5	6.1	15.4

These **values** may only be used when a good-sized movement from the bowels occurs regularly; in many diseases associated with diarrhœa this consumption suffers to a great extent. That of **fat** is much affected in icterus and atrophy of the pancreas, in severe anæmias, and in most diarrhœal diseases.

From the values obtained a **metabolic balance-sheet** is made something like the following:—

DISEASE: CARCINOMA OF THE STOMACH (AGE, 49 YEARS)

INCOME

Date	Body Weight, Pounds	Nourishment	N	Fat	Carbohydrates	Calories
I. 12	115	1500 g. milk	7.8	16.5	61.5	—
		85 g. bread	1.1	0.54	49.5	—
		40 g. butter	—	34.8	—	—
		4 eggs	4.0	19.6	—	—
Total			12.9	71.4	111.0	1474
I. 13	115	2000 g. milk	12.4	22.0	82.0	—
		110 g. bread	1.4	0.7	64.0	—
		40 g. butter	—	34.8	—	—
		4 eggs	4.0	19.6	—	—
Total			17.8	77.1	146.0	1763
Average			15.35	—	—	1618.5

EXPENDITURE

Date	Urine			Fæces		N	Total N
	Amount	Sp. Gr.	N	Moist	Dry		
I. 12	1350	1022	21.6	317	87.2	2.66	22.6
I. 13	1750	1015	23.4				24.4
Total							47.0
Average							23.5

Therefore the daily average is:—

Nitrogen income = 15.35
Nitrogen expenditure = 23.5

And the daily expenditure of 8.2 $\dot{N}$ = 241.1 g. of muscle tissue.

In this balance-sheet it is often necessary to estimate the N in reference to urea, or to albumin, or to muscle tissue. To render this estimate easy, the constant relations existing between them are here mentioned:—

Nitrogen : Urea :: 1 : 2.143
Nitrogen : Albumin :: 1 : 6.25
Nitrogen : Muscle tissue :: 1 : 29.4
Urea : Nitrogen :: 1 : 0.466
Urea : Albumin :: 1 : 2.9
Urea : Muscle tissue :: 1 : 13.71

The differential diagnosis of proteid metabolism is only called for in some special rare instances, as, for instance, in determining the malignant or benign nature of tumors.

The significance of these metabolic balance-sheets lies chiefly in the possibility of controlling by their study the *nutrition of the patient to the most exact degree*, and of always harmonizing the diet with each state of nutrition and body change.

In **diabetes mellitus** the constant attention paid to metabolism is of immediate significance in diagnosis and in treatment. Two varieties of this form of diabetes are recognized, either of which may become converted into the other:—

1. *The mild form*, in which sugar only appears in the urine after the *ingestion of carbohydrates;* the sugar excretion is larger or smaller according as the amount of carbohydrates ingested is greater or less.
2. *The severe form*, in which sugar is contained in the urine even after many days have passed in which *no carbohydrates* at all have been taken as food.

Only a carefully prepared metabolic balance-sheet will give an accurate knowledge of diabetics and their treatment.

As an example, I give a balance-sheet of a mild case of diabetes.

INCOME

Date	Body Weight in Pounds	Nourishment	N	Fat	Carbo-hydrates	Calories
III. 15	115	1 litre milk	6.2	31.0	45.0	—
	—	10 eggs	10.5	49.0	—	—
	—	120 g. butter	—	104.4	—	—
	—	125 g. meat	4.2	1.1	—	—
	—	60 g. bread	0.8	0.4	35.0	—
Total			21.2	185.9	80.0	2600

EXPENDITURE

Date	Urine				Fæces		N	Total N
	Amount	Sp. Gr.	N	Sugar	Moist	Dry		
III. 15	2800	1022	18.8	33.6 (1.2 %)	238	47.6	1.7	20.5

Therefore of 80 g. of carbohydrates, 46.4 g. are consumed regularly, and 33.6 g. are excreted in the urine unconsumed. The excretion of nitrogen is somewhat less than the amount taken in.

Symptoms of a Few of the Metabolic Disorders

Diabetes mellitus. — The diagnosis is settled by a positive reaction to the test for sugar. It depends whether an examination of the urine is thought of at the right time. The following signs should induce the physician to make a urinary examination for sugar: diminution of physical and mental capacity, progressive cachexia, very much increased appetite (polyphagia), much increased thirst (polydipsia), large excretion of urine (polyuria); in certain cases there is a tendency to the production of boils (furunculosis), simple wounds heal with difficulty, severe pruritus is present, the presence of certain diseases of the eye (cataract, optic neuritis), and an early development of impotence occurs. Concerning the mild and severe forms, see p. 238: ferric chloride reaction, p. 200.

Gout is characterized by repeated short attacks of inflammation of the joints, especially of the metatarso-phalangeal joint of the large toe, although it also occurs in other joints. After the attacks deposits of uric acid often accumulate in the cartilages of the joint (gout nodules, tophi). Tophi also are deposited in the cartilage of the ear and in the skin, especially of the leg. After frequent attacks deformities of the joints ensue. Interstitial nephritis and atrophic kidney (gout of the kidneys) often develop. In the intervals between the attacks there are many nervous disturbances (intervallary symptoms) and often inflammatory affections of the internal organs (visceral gout).

Diseases of the Thyreoid Gland

Basedow's disease is characterized by struma, exophthalmus, tachycardia (often associated with dilatation of the left ventricle and a systolic murmur), tremor of the fingers, and nervous irritability. There is often cachexia.

Myxœdema (cachexia strumipriva) is characterized by absence of the thyreoid gland, a swelling of the skin of the entire body, gradual failure of the physical and psychical powers, loss of the hair, and progressive cachexia.

CHAPTER XI

DIAGNOSIS OF THE DISEASES OF THE BLOOD

As far as the history is concerned an inquiry into the hygienic conditions of the patient, his habits, and his occupation is of value, because psychical excitement, troubles, and worry often lead to anæmia. All conditions which lead to chronic *loss of blood* may be regarded as direct causes of anæmia: ulcer of the stomach and of the intestine, uterine myomata, profuse menstruation, certain intestinal parasites (the *Anchylostomum duodenale* and *Bothriocephalus latus*); finally, anæmia is produced by all those severe *disturbances of digestion* which are occasioned by atrophy of the mucous membrane of the stomach and of the intestines, chronic intestinal catarrh, and continued attacks of diarrhœa. Every severe injury to the organism, as well as every disease of long standing (syphilis, for example), may lead to true anæmia. However, many blood diseases develop without a discoverable etiological cause; in such cases the history should concern itself in part in inquiring carefully about the indefinite symptoms which usually precede the full development of the disorder, such as malaise, loss of energy, disturbed sleep, headache, cardiac palpitation, dyspepsia often, etc.

A diagnosis is directed to a disease of the blood when there is an intense *pallor* of the skin and the mucous membranes (see p. 7), associated with physical weakness.

As has been already mentioned among general symptoms, a blood disease may be *secondary*, that is, caused by a severe visceral disease which leads to bodily wasting, such as tuberculosis, carcinoma, amyloid degeneration, etc. Only when such a disease can be excluded may a diagnosis of a specific disease of the blood be considered; *an examination of the blood* will then substantiate it.

It is often possible to diagnose an essential (not a secondary) disease of the blood from the peculiar color of the skin. The skin in pernicious anæmia is waxy yellow, often having a greenish tinge; this color is wholly characteristic.

The examination of the blood should concern itself with —
1. The macroscopic appearance of the blood.
2. The ordinary microscopic examination.
3. The counting of the corpuscles.
4. The measuring of the corpuscles.
5. The preparation of *stained* blood specimens.
6. The determination of the amount of hæmoglobin.

The scientific analysis of blood diseases includes in addition, among other things, the *reaction*, the amount of *carbonic acid* it contains, and the investigation of its *metabolism* (see previous chapter). A *spectroscopic* examination is necessary for diagnostic purposes in many cases of poisoning.

For the purposes of examination blood should be obtained by pricking the finger tip or the lobe of the ear, which has been previously washed and dried. The prick should be made with a sharp needle or, better, with a vaccination lancet; it should be deep enough that the blood will exude of its own accord; pressure must not be used; the first drop is wiped away, the second alone is to be examined.

1. The Macroscopic Examination of the Blood

includes the *color*, which normally should be of a bright red, and which in disease becomes paler and approaches a white. The rapidity with which it coagulates is also to be observed. If we accustom ourselves to prick as uniformly as possible, then the abundant or scanty amount exuding from the prick will furnish some conclusion as to the amount of blood. This can hardly be considered, though, in a differential diagnosis.

2. The Examination of the Fresh Blood Drop under the Microscope

A drop of blood is put on a slide, the cover glass carefully added; it is best to avoid evaporation and consequent drying by covering the edges of the cover glass with some warm paraffine. In the examination the following features should be observed: —

a. The shape of the red blood corpuscles; normally they have the shape of a disc with a central depression. The shape is unchanged in chlorosis and anæmia. In all severe anæmias changed forms appear, such as : *poikilocytes* (Fig. 40), which are club-shaped, pear-shaped, biscuit-shaped, or kidney-shaped red blood corpuscles; *microcytes,* red blood corpuscles much smaller than the ordinary red corpuscle; and *macrocytes,* which are decidedly larger than these.

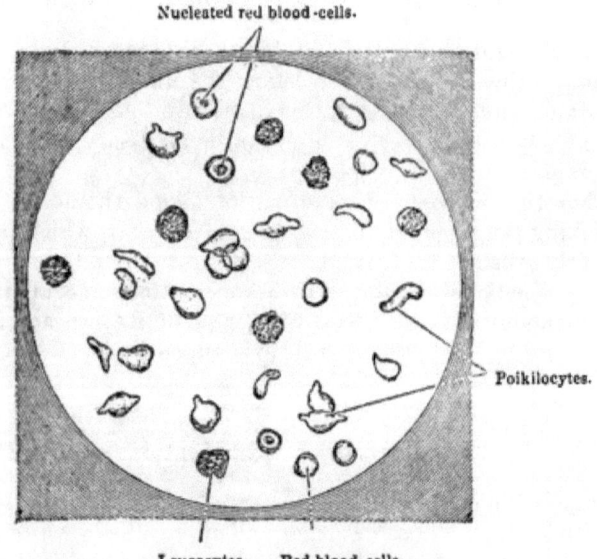

FIG. 40. — THE BLOOD IN PERNICIOUS ANÆMIA.

b. The tendency to the rouleaux formation; this tendency is absent in all conditions of extensive diminution in the number of red corpuscles, therefore in all severe anæmias.

c. The number of the red blood corpuscles; although this can only be positively determined by a counting apparatus, yet after practice in preparing specimens uniformly we can reach the position to tell whether the number is essentially diminished by the ordinary microscopic appearance. The increase of erythrocytes is the sign of anæmia.

d. The color of the red blood corpuscles, normally yellowish-red, is more or less pale in many diseases, especially in chlorosis.

e. The number of the white blood corpuscles and their relation to the red. Normally, 1 white exists to every 300 red blood corpuscles, or, in other words, in a microscopic field with open diaphragm and a high objective (Leitz 7), 3 to 5 white corpuscles will be present.

The abundant presence of leucocytes in a visual field (over 10) is an important sign of disease. Moderate increase of the leucocytes (1 white to 100 red) is called a *hyperleucocytosis* (see below). Very great increase of the leucocytes is the sign of *leucæmia* (1 to 50 and up to 1 to 2). These results should be controlled by the counting apparatus.

With practice with the ordinary microscopical examination we may be able to recognize, for instance, the nucleated red blood corpuscles as well as the variations in the white corpuscles. Yet these features are more easily recognized in examining stained preparations.

3. Counting the Blood Corpuscles

To count, the *Thoma-Zeiss* counting apparatus is used. It consists of a glass capillary tube that is blown out near the upper extremity into a bulb, which serves to draw up and to dilute the blood, and of a counting chamber. The blood is drawn up into the graduated tube to the mark 0.5 (or to 1), the point of the tube is carefully wiped, and then some of a 3 per cent. solution of sodium chloride (salt) is drawn into the tube until the whole contents of the tube reach the mark 101. The fluid in the tube is then thoroughly mixed by shaking (this is facilitated by the little glass ball which is found in the bulb). After this the mixture is placed in the counting chamber, which is exactly 0.1 mm. deep and the bottom of which is divided into microscopical squares. The space over each square is exactly $\frac{1}{4000}$ c.c. When placing the cover glass over the counting chamber, care should be taken to prevent the formation of air-bubbles. A large number of these squares should be counted, every 16 of which are marked off by darker lines, and in this way we obtain the average number of blood corpuscles lying in each square. This average is then multiplied by 800,000 (if the blood was sucked up to the mark 1, it is only multiplied by 400,000), for the reason that the blood was diluted 100 times and the space over each square is $\frac{1}{4000}$ c.c.; the result will give the number of red blood corpuscles in every cubic millimetre of blood.

The counting of the white corpuscles is performed in exactly the same way, although the blood is diluted in a separate pipette only 10 times. It is of advantage to add some methyl-violet to the diluting fluid, for the leucocytes take up this stain and appear much more prominently than otherwise. The destruction of the red blood corpuscles (by diluting the blood with 1 per cent. of acetic acid) is recommended in counting the white corpuscles.

In healthy men the number of the red blood corpuscles is 5,000,000, in women between 4,000,000 and 5,000,000, in every cubic millimetre. In *chlorosis* the number is not at all or but little changed, but in all anæmias it is much diminished, even down to 500,000; in severe leucæmia there is also a diminution of the number of red blood corpuscles.

The number of white blood corpuscles in health is 5000 to 8000 in every cubic millimetre. An increase in the number (*hyperleucocytosis*) occurs physiologically during the digestion of the *albuminoids* (10,000 to 20,000) and occurs pathologically in many infectious and cachectic diseases, *e.g.*, pneumonia and carcinoma (see p. 250). Only an increase to over 50,000 in a cubic millimetre would justify a diagnosis of *leucæmia;* this diagnosis will be probable if with an existing hyperleucocytosis the number of leucocytes increases in a short time.

4. Measuring the Blood Corpuscles

The size of the red blood corpuscles may be well estimated and macrocytes and microcytes recognized with sufficient distinctness. For a careful examination a *micrometer* eyepiece should be used. The red blood corpuscles of the healthy vary in size from 6.5 to 9.4 μ (microns), on the average they are 7.6 μ; they are of equal size in the same individual. Macrocytes are those whose size is 10 to 12 μ, gigantocytes 12 to 15 μ. Their presence denotes a grave anæmia. The size of the leucocytes varies extremely.

5. The Preparation and Microscopical Appearance of Stained Specimens

(According to *Ehrlich*)

The drop of blood is taken up from the finger tip directly on to the cleansed cover glass, which is placed lightly on another clean cover glass in such a manner that the edges do not coincide and then the glasses are drawn apart without exerting pressure, which must be avoided. Care should be taken that the fingers do not touch the cover glasses, because the warmth and moisture of the skin will change the very sensitive blood corpuscles. The blood has now been finely spread in a thin layer on both cover glasses. The spreads are now dried by the air and thereafter fixed by heat. The heating must be gradual, for which purpose the cover glasses are placed in a hot-air oven or on a copper plate; the latter is heated at one end to 120° C. and kept at this temperature for 2 hours. After they have cooled they are ready for staining.

The dye most frequently used is the eosin-hæmatoxylin solution (hæmatoxylin, 2; alcohol, glycerine, and distilled water, $\bar{a}\bar{a}$ 100; glacial acetic acid, 10, to which alum in excess is added; the solution should stand several weeks, then a small quantity of eosin is added to it). The spreads are put in this solution and kept there for 30 minutes, when they are removed and washed with water; the red blood corpuscles will appear red, the nuclei of the white and those of the red are deeply stained bluish-black, and the eosinophile granulations (see below), red; the protoplasm of the white blood cells is almost unstained, having only a pale-red tint.

Ehrlich's three-color mixture (see p. 144) is very serviceable for staining the blood. This stains the nuclei a greenish-blue, the eosinophile granulations red, the red blood cells orange.

Beautiful results are obtained with the eosin-nigrosin-aurantia-glycerine mixture. (To 1 volume of glycerine saturated with aurantia, add 1 to 2 volumes of glycerine; shake the mixture thoroughly and then add eosin and nigrosin in excess; saturation results after long-continued shaking.) The hæmoglobin takes up the yellowish-red tint of the aurantia, all nuclei are gray or black, and the eosinophile granules are red.

The stained preparations are best examined with an oil immersion lens and with open diaphragm.

In a stained specimen we recognize: —

1. The nucleated red blood corpuscles; these are always a sign of severe blood disease; they are present in many anæmias, and occur less seldom in leucæmia; nucleated megalocytes and gigantocytes indicate very grave anæmia. Still in spite of their presence, improvement, even a cure, may result.

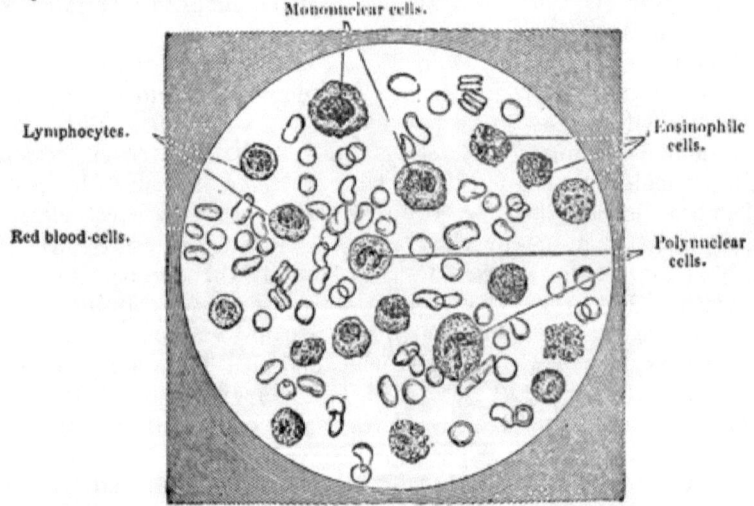

FIG. 41. — SCHEMATIC REPRESENTATION OF VARIOUS KINDS OF LEUCOCYTES.

2. The different forms of the *white* blood corpuscles: —

    a. Lymphocytes, of varying size; usually somewhat larger than a red blood corpuscle, possessing a round nucleus and small protoplasmic body; they arise from the lymphatic glands, and a marked increase of them indicates *lymphatic leucæmia.*

    b. Mononuclear leucocytes are much larger than the red corpuscles, and have a large ovoid nucleus and a large protoplasmic body; from them are developed the

    c. Polynuclear leucocytes, which have a polymorphous nucleus, and form the mass of the leucocytes;

they are chiefly present in pus. Mono- and polynuclear cells are increased in *splenic* and in *myelogenic leucæmia*.

d. *Eosinophile cells*, large, round, and nucleated; they are characterized by the glistening granules in the protoplasmic body of the cell; these granules are deeply stained by eosin. These cells arise in the bone-marrow, are seldom found in normal blood, and when abundantly present justify the conclusion that there is an implication of the marrow of the bone.

6. The Determination of the Amount of Hæmoglobin in the Blood

The amount of hæmoglobin in the blood is determined with sufficient accuracy by means of *Fleischl's hæmometer*. The color of the blood diluted with water is compared with a wedge of glass which is colored purplish-red. The blood is sucked up in a capillary tube of known dimensions, dissolved in water in the one compartment of a glass box which is partitioned into two divisions; under the other compartment, which is filled with water alone, the colored glass wedge, whose color increases in intensity as the wedge increases in depth, moves across. The wedge is provided with a scale, empirically determined in such a manner that the 100 mark of its color corresponds to the normal hæmoglobin in the blood. Under the glass bottom of the box a light is reflected by a gypsum plate; should then the color be alike in both compartments, the number indicated on the scale at this stage will denote the hæmoglobin contents of the blood. The limits of error in this instrument are within 15 per cent. *Gower's* hæmoglobinometer determines the amount of the coloring matter of the blood by comparing a specimen of diluted blood with some artificially colored glass rods. Pretty accurate results are obtained by the expert who uses photometric spectral analysis suggested by *Vierordt;* but its use is extremely difficult.

The amount of hæmoglobin in the blood is much diminished in chlorosis, while the number of red corpuscles is not at all affected. In the other anæmias a diminution in the amount of hæmoglobin will coincide with a decrease in the number of red blood corpuscles.

The absolute amount of hæmoglobin is 13 to 15 g. to every 100 c.c. of blood, in women usually a little less than in men.— Hæmoglobin is converted into *albumin* and *hæmatin*. Hæmatin with hydrochloric acid produces *hæmin*, which form in beautiful crystals (*Teichmann's*), by whose presence the slightest trace of blood may be recognized.

Teichmann's blood test.— A small quantity of dried blood, to which 1 to 2 drops of glacial acetic acid and a small crystal of common salt are added, is heated in a glass dish over a full flame until it boils; it is then permitted slowly to evaporate, when numerous brownish-yellow needles and crystals of hæmin will be formed.

The reaction of the blood is alkaline; the alkalinity is diminished in severe anæmias, in fever, in severe diabetes, and in the emaciation stage of carcinoma.

The reaction of the blood may not be determined simply with litmus, etc., on account of the individual color of the blood and chiefly because the blood contains different acids and bases in varying degrees of saturation. In one sense the alkalinity of the blood may be reckoned from the amount of *carbonic acid* the blood contains, for there is an approximately standard relation between the alkalescence and the amount of CO_2.

The determination of the *specific gravity* has thus far been of little use for practical diagnostic purposes; in health it varies between 1045 and 1075.

The *spectroscopic* examination of the blood is of importance in diagnosing carbonic oxide intoxication.

Normal blood strongly diluted with water will show the absorption bands of *oxyhæmoglobin* in the yellow and green (between the Frauenhofer lines of D and E). On adding a few drops of a solution of ammonium sulphide, the two lines will vanish and a single line (between D and E) of reduced hæmoglobin will appear.

The bright red blood of carbonic oxide intoxication shows likewise, when viewed in the spectroscope, two lines between D and E, though these appear closer to each other than the oxyhæmoglobin lines. On the addition of ammonium sulphide the lines of the CO hæmoglobin do not disappear.

In cases of poisoning with *potassium chlorate*, anilin, antifebrin, phenacetin, etc., the color of the blood is like chocolate, and in the spectroscope, in addition to the two oxyhæmoglobin lines, an absorption line is observed in the red which belongs to *methæmoglobin*. On adding some ammonium sulphide, all three lines disappear and the one line of reduced hæmoglobin puts in an appearance.

CHIEF SYMPTOMS OF THE MOST IMPORTANT DISEASES OF THE BLOOD

Chlorosis occurs in young persons, especially in young girls; still sometimes it occurs in women, especially after parturition. It is characterized by pallor of the skin, intense fatigue, often by dyspepsia, palpitation, etc. The essential changes in the blood are a *great diminution* in the *amount of hæmoglobin* without an essential decrease in the number of red or an increase in the number of white blood cells. The prognosis is usually good.

Hyperleucocytosis is a temporary increase in the number of white blood corpuscles; it is a symptom of many inflammatory diseases (occurs principally in pneumonia, erysipelas, meningitis, and in the cachectic diseases, especially in *carcinoma*). In typhoid fever, malaria, glanders, and in many forms of sepsis hyperleucocytosis is absent. It occurs physiologically during digestion. Hyperleucocytosis may extend to an increase to 100,000 leucocytes in one cubic millimetre of blood. The characteristic differential point between it and *leucæmia* lies in the demonstration of the original disease and in the increase of the polynuclear leucocytes alone. The prognosis is dependent on the original disease producing it.

Leucæmia is characterized by a great increase in the number of white blood corpuscles, so that the relation may be from 1 to 60, to 1 to 2. It is to be differentiated from the primary stage of a leucocytosis in the more rapid increase of the leucocytes in leucæmia. The red cells are usually diminished in number, are often nucleated, and the amount of hæmoglobin is also diminished. The following forms exist, although *each may merge into the others*.

1. *Lymphatic leucæmia.* Swelling of all the lymph glands; the lymphocytes of the blood are increased.

There is a well-characterized form of **acute leucæmia** in which only the lymphocytes are increased. It runs a rapidly fatal course, accompanied by the clinical phenomena of hæmorrhagic diathesis, dyspnœa, enlargement of the lymph glands and spleen; in many organs leucæmic hyperplasia appears. The uric acid excretion is enormously increased.

2. *Myelogenic leucæmia.* There are many eosinophile as well as mononuclear leucocytes and nucleated red blood corpuscles.

3. *Splenic leucæmia.* The spleen is very large. The blood contains many eosinophile and mononuclear cells. A progressive cachexia characterizes all three forms, which terminate fatally.

Pseudo-leucæmia is the name given to the disease which has the clinical features of leucæmia, such as cachexia, lymphatic gland enlargements, and a splenic enlargement, but which runs its course *without the changes in the blood characteristic of true leucæmia*. The number of leucocytes is normal and there is little diminution of the red cells and of hæmoglobin.

Pseudo-leucæmia, presenting large swellings of the lymph glands, is called *Hodgkin's disease*.

Pernicious anæmia. The number of red blood cells is much diminished, even to 400,000 in each cubic millimeter. Poikilocytes, macrocytes, and microcytes are in abundance. There are nucleated red corpuscles and nucleated gigantocytes. The amount of hæmoglobin is relatively increased, while the number of leucocytes is normal or even diminished. The prognosis is most grave.

Secondary anæmia occurs in the course of severe dyspepsia, anchylostomiasis, carcinoma, phthisis, tertiary syphilis, malaria, amyloid degeneration, chronic intoxications, etc. There is a large decrease of the red blood cells; macrocytes and microcytes are present, but rarely are there any nucleated red cells; the amount of hæmoglobin is also diminished. The number of polynuclear leucocytes is *increased*. The prognosis of secondary anæmia depends on its causative disease; should we be successful in removing that, the anæmia may be cured. Secondary anæmia may go into pernicious anæmia; on this account in some cases a differential diagnosis may be extremely difficult.

CHAPTER XII

ANIMAL AND VEGETABLE PARASITES

I. Animal Parasites

The animal parasites which are found in or upon the human body are, in part, harmless organisms which infest the skin or the intestines and are without diagnostic significance; in part, however, they are productive of diseases, more or less severe, the treatment of which is dependent almost wholly upon a correct diagnosis.

In the following summary, the principal animal parasites are included: —

I. **Protozoa.**
    a. Rhizopods: Monads and Amœba coli.
    b. Sporozoa: Coccidia.
    c. Infusoria: Cercomonas intestinalis, Trichomonas intestinalis, Paramecium coli.

II. **Vermes** (Worms).
    a. Tape-worms (Cestodia).
        1. Tænia solium.
        2. Tænia mediocanellata or saginta.
        3. Bothriocephalus latus.
        4. Tænia nana.
        5. Tænia flavopunctata.
        6. Tænia cucumerina.
        7. Tænia echinococcus.
    b. Flukes (Trematoda).
        1. Distoma hepaticum.
        2. Distoma lanceolatum.
        3. Distoma hæmatobium.
    c. Thread worms or *Round worms* (*Nematoidea*).
        1. Ascaris lumbricoides.

2. Ascaris mystax.
3. Oxyuris vermicularis.
4. Anchylostoma duodenale.
5. Tricocephalus dispar.
6. Trichina spiralis.
7. Anguillula intestinalis.
8. Filaria sanguinis.

III. Arthrozoa.
1. Acarus scabiei.
2. Acarus folliculorum.
3. Pediculi.
4. Pulex irritans.

The **Protozoa** are practically without diagnostic significance. They are round, granular organisms, about $1\,\mu$ in length, some of the infusoria being a little larger, provided with ciliæ or flagellæ. They are found in healthy fæces, in the discharges of chronic diarrhœa, and sometimes in the normal vaginal secretions. Only the amœba coli is the causative factor of dysentery, and has therefore diagnostic significance.

Tape-worms

The tape-worms are principally intestinal parasites. As such they evoke a complex of dyspeptic, dysenteric, and nervous symptoms which are, in a measure, exceedingly painful and which disappear upon the expulsion of the parasite. The diagnosis of the presence of a tape-worm can be established only by the passage of segments (proglottides) of the worm.

Tape-worms consist of a *head* (scolex) and *joints* (proglottides). They reproduce by sexual alternation. The bisexual segments bud from the head, the impregnated eggs entering the stomach of some second animal, the *intermediate host*. Here the coverings of the eggs are digested and the embryo becomes free. It reaches the tissues of the intermediate host in the form of a *cysticercus*. If the cysticercus reaches the stomach of man with his food, a new tape-worm is developed.

Tænia solium (intermediate host, the pig) attains a size of from 2 to 3 M., its proglottides are from 9 to 10 mm long and from 6 to 7 mm. broad. The proglottides nearest the head are short and gradually increase in size as the tail is approached. The head has the size of a pin. Under the microscope may be seen four sucking discs, usually pigmented, and a proboscis or rostellum with from 25 to 30 hooklets of different sizes (Fig. 42). The segments have lateral sexual openings and a uterus with few branches. The eggs are oval, about 0.036 mm. long and 0.03 mm. broad, with a thick covering and radial striæ. In the interior of the ovule, the hooklets of the embryo are visible. The larvæ (*Cysticercus cellulosæ*) are the size of a pea and may be deposited in the organs of the body (mainly the skin, the muscles, the brain, the eye) by auto-infection, if they have entered the stomach.

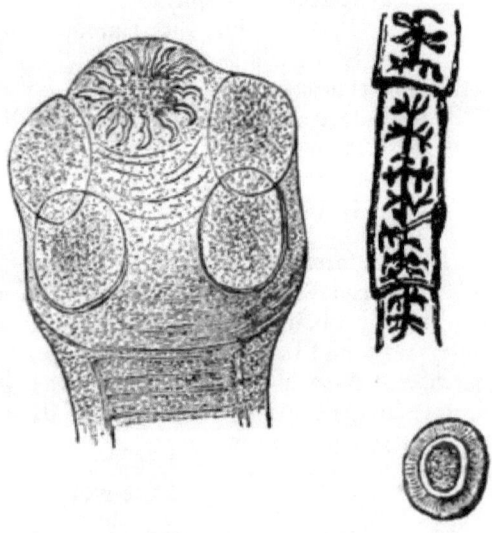

FIG. 42.—MICROSCOPIC PICTURE OF TÆNIA SOLIUM (HEAD, PROGLOTTIDES, EGG).

Tænia solium is the tape-worm most frequently found in the intestine. It is usually possible to distinguish this species with the naked eye or with a magnifying glass, by the delicacy and transparency of its segments and by the few ramifications (7 to 12) of its uterus.

The diagnosis of *cysticercus cellulosæ of the skin* is usually easy to make. Multiple movable tumors, from the size of a pea to that of a bean, are observed. Excision and examination makes the certainty of diagnosis more positive. The presence of a cysticercus in the *eye* may be demonstrated ophthalmoscopically. The diagnosis of cysticercus in the *brain* is made probable, if focal cerebral symptoms appear without demonstrable etiology, and if, at the same time, cysticerci are present in the skin or in the eye.

Tænia saginata or **mediocanellata** (intermediate host, cattle) are from 4 to 5 M. long. The head has no rostellum or circle of hooklets, but is provided with four very powerful sucking discs (Fig. 43). The proglottides are longer than those of *Tænia solium* and do not taper as they approach the head. The generative cloaca is situated at the sides and the uterus has many ramifications. The eggs are somewhat more oval than those of *Tænia solium*, but otherwise the resemblance is marked, although the hooklets of the embryo are not visible. The larvæ does not develop in the tissues of the human body.

Fig. 43.—Microscopic Picture of Tænia Saginata (Head, Proglottides, Egg).

The segments of a *Tænia mediocanellata* may be recognized with the naked eye; they are thicker and not so delicate as those of *Tænia solium* and the uterus has many more ramifications (from 15 to 20).

Bothriocephalus latus (intermediate host, various fishes, pike, salmon, etc.; geographical distribution limited mainly to the coast of the Baltic and Switzerland) is from 4 to 15 M. long. Its head is 2 mm. long, 1 mm. broad, club-shaped, and is provided in the median line with grooves which act as sucking discs (Fig. 44). The segments nearest the head are short and narrow, those further away almost quadrilateral. The uterus, when filled with eggs, is brown in color and presents a star-shaped form. The ovules are oval, 0.07 mm. long, 0.045 mm. broad, have a brownish shell, and are provided with a little cover.

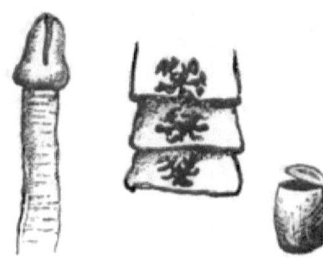

Fig. 44.—Microscopic Picture of Bothriocephalus Latus (Head, Proglottides, Egg).

The diagnosis of *Bothriocephalus latus* is of importance, since its presence may evoke a profound *anæmia*, which disappears upon the expulsion of the tape-worm. The proglottides of bothriocephalus may be recognized by their brown color and by the rosette-shaped uterus.

Tænia nana is from 10 to 15 mm. long, 0.5 mm. broad. The diameter of the head, which has a rostellum and sucking discs, is 0.3 mm. The segments are short, 4 times as broad as they are long. The uterus is oblong in shape. The ovules measure from 0.03 to 0.04 mm. in diameter; they are surrounded by a double membrane without a striated shell. In the interior of the egg, the embryo with its hooklets is visible. From 4000 to 5000 of these tape-worms may infest the intestines simultaneously. Tænia nana has been seen only in southern countries (Italy, Egypt), and is said to call forth severe psychical and nervous disturbances.

Tænia flavopunctata and **Taenia cucumerina** are extraordinarily rare.

Tænia echinococcus is found in the human body only in its larval stage.

The tape-worm itself exists only in the intestines of dogs. It is 4 mm. long. The head possesses a circle of from 20 to 30 hooklets. The embryo reaches the human stomach and intestine and becomes encysted. The cyst-wall consists of two layers, an outer, finely lamellated layer, the cuticula, formed of chitinous material, and an inner layer, the parenchymatous, which contains muscle fibres and blood-vessels. The scolex or head develops in the parenchymatous layer and is provided with hooklets and sucking discs. The echinococcus cyst may be *unilocular* or secondary or *daughter cysts* may develop within it, or it may consist of a mass of small cavities, filled with a colloid fluid, the walls of which may show concentrically arranged layers. This is known as a *multilocular echinococcus cyst*.

Echinococcus cysts have their principal seat in the liver, but are found, less often, in the lungs, the brain, or the heart.

The symptoms evoked are those of a large cyst the nature of which is determined by aspiration and microscopic examination of the aspirated material. Sometimes, characteristic *membrane* and *hooks* (Fig. 45) are seen, or *chemical* examination of the fluid discloses some of its peculiar properties.

The fluid of an echinococcus cyst is usually clear, of a specific gravity of from 1008 to 1013. It contains *little* or *no* albumin, but *sodium chloride* is present in large quantities, and frequently grape-sugar and succinic acid are found.

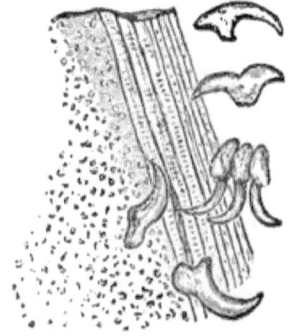

Fig. 45. — Echinococcus Membrane and Hooklets.

The presence of succinic acid may be determined as follows: the fluid is heated over a water bath, is made acid by the addition of hydrochloric acid, and is then shaken up with ether. The ether is evaporated and if succinic acid is present, a crystalline jelly remains, which is to be dissolved in water. With ferric chloride, succinic acid gives a rust-colored, colloid precipitate which on heating in a test-tube gives off a vapor exceedingly irritating to the mucous membranes.

Flukes (Trematoda)

Distoma hepaticum is of a leaf-like form, has a short, globular head, and attains a length of 28 mm. The ovules are oval, 0.13 mm. long, 0.08 mm. broad, and are provided with a cover (Fig. 46). This parasite is rarely found in the biliary passages in man, and its eggs have been occasionally found in the intestines. Its diagnostic importance is insignificant, but the possibility of its confusion with eggs of important diagnostic worth must be mentioned.

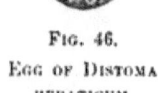

Fig. 46. Egg of Distoma hepaticum.

Distoma haematobium occurs only in the tropics. Its residence is in the portal system of veins and in the veins of the bladder and of the rectum. It causes diarrhœa, hæmaturia, and ulcerations of the mucous membranes.

Fig. 47. Egg of Distoma haematobium.

The male worm has a length of from 12 to 24 mm., the female from 16 to 19 mm. The abdomen of the male has a sulcus opening downward, in which the female is carried. The eggs (Fig. 47) are found in the lungs, the liver, and the bladder, are 0.12 mm. long and 0.04 mm. broad, and are provided at the end or side with a thorn-like projection.

Distoma lanceolatum is a lancet-shaped worm, 7 to 8 mm. long, 2 to 3.5 mm. broad, smaller than but similar to *Distoma hepaticum*. The parasite is very rarely found in the gall-bladder or gall-passages, and the eggs are still more rarely seen in the fæces. Its diagnostic importance is slight.

Thread Worms or Round Worms (Nematoidea)

Ascaris lumbricoides is the common thread worm. The male attains a length of 25 cm., the female of 50 cm. It appears in large numbers in the human small intestine. In general it is a harmless parasite, but sometimes reflex convulsions in children are attributed to it. The eggs are found in the fæces in large numbers. They are round, yellowish-brown, of a diameter of 0.06 mm. In the fresh state, they are surrounded by a crenated, albuminous covering (Fig. 48), which in turn surrounds a thick, concentric, striped shell which contains a granular material.

FIG. 48.
EGG OF ASCARIS LUMBRICOIDES.

Oxyuris vermicularis, thread worm or pin worm.

FIG. 49.
EGG OF OXYURIS VERMICULARIS.

The male is 4 mm., the female 10 mm. in length. The parasites are found in the intestines in large quantities. The eggs are 0.05 mm. long and 0.02 mm. broad, are oval and more pointed at one end than at the other. They have an edge with a double contour (Fig. 49). The worm often leaves the intestine, and, remaining in the neighborhood of the anus, evokes a most troublesome itching.

Anchylostoma duodenale is diagnostically of the greatest importance, because, by its constant sucking of blood from the walls of the intestine, it produces a profound anæmia which may resemble, in its clinical course, a case of pernicious anæmia.

Anchylostoma is seen in brickmakers, miners, and workers in tunnels, and when an anæmia presents itself in laborers of this class, the fæces must always be examined for anchylostoma. As long as no anthelmintic is administered, only the *eggs* are found in the stools.

The male is from 8 to 12 mm., the female from 10 to 18 mm. long. The male has a three-lobed tail, the female a pointed, conical tail. The cephalic end is provided with a bell-shaped mouth-capsule which contains four claw-like teeth. The eggs (Fig. 50) are 0.05 mm. long and 0.03 mm. broad. They have a smooth surface and in the interior several *segmentation bodies* are visible. If the ovules are not absolutely recognizable at first, the fæces may be allowed to stand in a warm place for 2 or 3 days, when microscopic examination will disclose in anchylostoma eggs a decided increase in the segmenting process; or the patient may be given an anthelmintic, such as the extract of felix mas, in order to establish the diagnosis by the appearance in the fæces of the parasite itself.

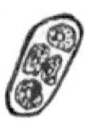

FIG. 50.
EGG OF ANCHYLOSTOMA DUODENALE.

Trichocephalus dispar (Fig. 51) is found in the large intestine, and is not of great diagnostic importance. The male is 4 cm., the female 5 cm. long. The eggs, 0.06 mm. long and 0.02 mm. broad, appear in large numbers in the stools, are brown in color, and are closed at each pole by a shining cover.

FIG. 51.
EGG OF TRICHOCEPHALUS DISPAR.

Trichina spiralis is found in the human body in the *muscles* and in the *intestines*. By the eating of improperly cooked, trichinous pork, the trichinæ reach the human stomach and intestine. Here the capsule containing them is dissolved and males (1.3 mm. long) and females (3 mm. long) become free, and multiply. In the course of from 5 to 7 days, the young trichinæ work their way through the intestinal wall, enter the blood-current, and are carried into the muscles, where they may become encapsulated (Fig. 52). The diagnosis of trichinæ is established by the appearance of the parasite in the fæces after the administration of anthelmintics — although this is rarely accomplished — or by the finding of trichinæ in the muscles. The symptoms of inspection by trichinæ depend upon the stage of the disease. If the parasites are still in the intestines, the symptoms of a gastro-enteritis present themselves; if they have reached the muscles, multiple abscesses in these organs appear.

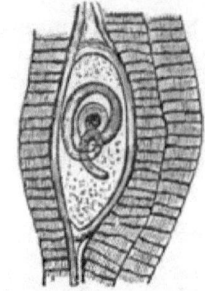

FIG. 52.

TRICHINÆ IN MUSCLE.

Anguillula intestinalis (Rhabdonema strongyloides, *Leuckart*) are 2.25 mm. long, with rounded, obliquely striped bodies, and appear in large numbers in the small intestine. The eggs bear a striking resemblance to those of anchylostoma duodenale, and a differentiation between the eggs of these species may, occasionally, be of importance. They are not known to exert any injurious influence.

Filaria sanguinis occurs chiefly in the tropics, evoking *hæmaturia* and *chyluria*. Great numbers of the embryos circulate in the blood. The parasite is an auto-mobile, delicate worm, surrounded by a thin membrane. It is 0.35 mm. long and about as broad as a red blood-cell. In the sediment of the urine the embryos may be found in abundance.

Filaria medinensis, also a tropical parasite, is a very long (80 cm.), very narrow worm (about 1 mm. broad). It produces a severe furunculosis.

Arthrozoa

The *head-louse* (Pediculus capitis), the *body-louse* (Pediculus vestimenti or Corposis humani), and the *crab-louse* (Pediculus pubis) must be given diagnostic consideration, since through their bites *eczema* and *excoriations* may be evoked, which may be confused with other skin diseases, and hence be wrongly treated.

The *flea* (Pulex irritans) and the *bed-bug* (Acanthia lecticularis) must be mentioned in this connection, since the flea-bite bears some resemblance to patechiæ, and may occasionally mislead to a diagnosis of purpura. The wheals arising from bed-bug bites look something like roseola.

Acarus scabiei, the itch-mite, is the etiological factor of the itch or scabies, which is recognized by the characteristic cuniculi made in the skin by the female, and the accompanying eczema. The male itch-mite is 0.2 mm. long, 0.35 mm. broad. The female is 0.35 mm. long, 0.5 mm. in width. Examined microscopically, the itch-mite bears some resemblance to a turtle with a conical proboscis and eight legs.

Acarus folliculorum, the pimple-mite, is found among the contents of hair-follicles (comedones). It is of worm-like form, 0.02 mm. broad, 0.1 mm. in length.

II. VEGETABLE PARASITES

1. Moulds and Fungi

The moulds are flowerless plants (cryptogamous) without stems or leaves, with simple foliage (thallophytes). The foliage consists of simple cells without nuclei and devoid of chlorophyll. They *never multiply by fission* but by the formation of long chains of cells (*hyphæ*). By the interbranching of these chains, a dense basket-work may arise (*mycelium*). Single hyphæ show peculiarities of growth, the *fruits* of the plants developing upon them. These fruits are called *spores* or conidia. According to the manner in which the fruit-bearing hyphæ develop from the mycelium and the manner in which these in turn form the conidia, the schizomycetes are divided into different groups: *mucorinæ, aspergillæ, penicillia,* etc.

The **saccharomycetes** do not form hyphæ nor mycelia; they consist simply of single cells without nuclei or chlorophyll. They multiply by *gemmation*. A bud appears on the surface of the mother-cell, which increases in size and finally detaches itself. Frequently, large masses of these cells cling together and form *colonies*.

There are **intermediate forms** between the moulds and fungi, which, under some conditions of nutrition, form hyphæ, and under other circumstances grow in colonies. The principal member of this group is the *Saccharomyces* or *Oidium albicans* (Soor fungus).

Achorion Schönleinii, the parasite of *favus*, was the first of the vegetable parasites of man to be recognized.

Trichophyton tonsurans is the fungus of *herpes tonsurans* and *parasitic sycosis*.

Both of these fungi have mycelia with many branches, with distinctly jointed hyphæ. In the fungus of favus the branches usually stand at right angles.

These fungi may be raised in characteristic pure cultures. A typical favus or herpes may be brought about by inoculation in the skin.

Microsporon furfur is the parasite of *Pityriasis versicolor*. The proof of the presence of this fungus is of diagnostic importance, since the yellowish scales of pityriasis — mostly seen in the cachectic diseases, particularly phthisis — may be easily taken for an actual pigmentation of the skin. The scales of pityriasis are easily removed and show under the microscope, particularly well upon the addition of a few drops of caustic potash, an entangled mycelium with heaps of shining conidia.

Aspergillus and *mucorina* are occasionally found in the external auditory passages, in the nostrils, and in the nasopharynx. The diseases they call forth, however, are evoked simply in a mechanical manner by their presence. The growth of fungi in the lungs (pneumonomycosis aspergillina) is usually a secondary deposit in necrotic tissue or cavities already existing.

Saccharomycetes often appear in the fermenting contents of the stomach, mainly in cases of dilatation, chronic catarrh, and carcinoma. By the action of this fungus sugar is split up into alcohol and carbonic acid gas.

Thrush fungus (*Saccharomyces* or *Oidium albicans*) has the power of producing necrosis of pavement epithelium, and is the cause of the grayish-white membranous deposits on the mucous membrane of the mouths of poorly nourished children and profoundly sick people. These membranes may arise in other organs of the body provided with pavement epithelium, as the œsophagus and vagina. In media rich in sugar and acid in reaction, this parasite grows in pure culture, as in the stomach; in alkaline media, as in the mouth, it grows with abundant hyphæ and conidia.

2. Schizomycetes (Bacteria)

In the category of these minute beings (micro-organisms) belong the producers of the infectious diseases. In the cases of some of the infectious diseases, the clinical demonstration of the specific micro-organism is indispensable.

Biology of Bacteria

The bacteria constitute the lowest plane of vegetable life. They appear in the following forms:—

1. *Micrococci* or *round bacteria* are arranged in chains (*streptococci*), or in grape-like bunches (*staphylococci*), or in pairs (*diplococci*).
2. *Bacilli* or *rod-shaped bacteria* may appear with curves (*vibriones* or *comma bacilli*), or they may grow into long chains (*leptothrix*).
3. *Spirilla*, screw form.

A dense mass of bacteria, bacilli, or cocci is known as a *zooglœa mass*.

Bacteria multiply by successive fission. Many bacteria, however, increase by *spores*. In the mother bacterium a granular, strongly refracting area becomes differentiated, is freed, and grows into a new bacterium. The spores represent the *permanent form* of the bacteria, which die very soon upon the application of a moderate degree of heat (50° to 60° C.) or under the influence of a somewhat concentrated antiseptic solution (3 per cent. solution of carbolic acid). The spores, however, are very resistant to every external influence, and can be killed with certainty only after being subjected for half an hour to the influence of boiling water, or for 3 hours to a dry heat of 110° C. The spores are not rendered innocuous with certainty by the usual dilutions of antiseptic solutions. *Pathogenic* and *non-pathogenic* bacteria are distinguished. The latter do not develop in the human body, but thrive on dead material (as *saprophytes*), producing *fermentation* and *putrefaction*.

Pathogenic bacteria thrive in the bodies of man and the lower animals, producing the *infectious diseases;* some of them, however, like the anthrax bacillus, can live on dead material. This variety is called *ectogenous* or *facultative*.

The Demonstration of Bacteria

For clinical purposes the main examinations of bacteria are confined to pus, sputum, aspirated fluids, fæces, and blood; for such study the mounting and staining of *dry specimens* is sufficiently satisfactory. In some cases, however, this method does not suffice, and the preparation of a pure culture (*Koch*) or animal inoculation must be resorted to.

The preparation of a dry specimen. A small particle of the substance to be examined is placed upon a perfectly clean *cover glass*, upon which another cover glass is then placed with great care. The two cover glasses are then repeatedly drawn over each other in order to get as fine a layer as possible on either glass. It is then allowed to dry as thoroughly as possible *in the air*, the prepared surface, of course, lying uppermost. The cover glass is then seized with a forceps and is passed two or three times horizontally through the flame of a Bunsen burner or an alcohol lamp. This is for the purpose of coagulating the albumin present. The specimen may now be subjected to the staining fluid.

If one wishes to make a hasty examination, the substance to be examined may be transferred at once to a *slide*, which, after heating in the flame, may be further treated the same as the cover glass.

The staining of a dry specimen. Alcoholic solutions of the *basic aniline dyes* must be kept at hand. The main ones are Bismarck brown, methylene blue, methyl-violet or gentian-violet, fuchsin (red), and malachite (green). To prepare these solutions, the crystalline powder of the dyes is dissolved in alcohol in excess, thoroughly shaken, allowed to stand several hours, and filtered. Four or five drops of the saturated alcoholic solution are placed in a watch-glass full of distilled water. The dry specimen is placed into this solution, the prepared side *down*, for from 2 to 4 minutes. It is then washed in water, dried between two pieces of filter paper, laid upon the slide in oil of cloves or Canada balsam, and is examined under the microscope. An *oil immersion* objective, an *open* diaphragm, and an *Abbé* condenser are essential for a proper bacteriological examination.

In order to stain rapidly, a concentrated watery solution may be dropped directly upon the specimen. Slide preparations are always treated in this manner, the staining fluid being washed off with water. They are microscopically examined without cover glasses.

The method just described answers ordinary clinical demands. The aniline dyes stain micro-organisms and the nuclei of the cells intensely; the protoplasm of the cell is usually stained very weakly.

By staining bacteria by *Gram's* method, they can be *isolated*, so far as coloring is concerned, from the tissues in which they lie. This method consists in placing the cover glass preparation in an aniline-gentian-violet solution for 3 minutes (see below, tubercle bacillus) and then for 1 minute in a solution of iodine and iodide of potassium (iodine, 1; iodide of potassium, 2; distilled water, 300). Decolorization is effected by repeated washing in alcohol. The bacteria appear on a colorless background, stained bluish-black. If it is desired, the nuclei of the cells of the tissues may be stained with some contrasting aniline color, as Bismarck brown or eosin.

The Staining of Tubercle Bacilli

1. *Ehrlich's method.* As described above, a dry specimen is prepared from the sputum. The particle selected should be taken from a purulent or cheesy part of the sputum. An easy way to find such particles is to pour the sputum upon a blackened plate or a piece of smoked glass.

The staining fluid is aniline-gentian-violet, and is prepared as follows: a saturated solution of aniline oil is made with 10 times its volume of water, and is filtered. An alcoholic solution of gentian-violet is added drop by drop to a watch-glass full of the clear aniline solution until a shining membrane appears on the surface.

The specimen, prepared side down, is placed in this solution and warmed for 10 minutes over a flame. The cover glass is then removed, washed with water and then in a 25 per cent. diluted nitric acid solution until it is colorless. Owing to its tenacity for aniline dyes, the tubercle bacillus alone is now stained, the other bacteria present having given up their coloring to the acid. To stain the tissues, the cover glass is laid in a solution of Bismarck brown for from 2 to 3 minutes. It is again washed in water and dried.

The tubercle bacilli are stained violet, the nuclei brown.

Instead of gentian-violet, one may add an alcoholic solution of fuchsin to the aniline solution, and the nuclei may be stained with methylene blue or malachite. In this event, the bacilli will be red, the nuclei blue or green.

2. *Fränkel-Gabbet's rapid method.* The following solutions must be prepared: —

   A. Fuchsin, 1
   Alcohol, 10
   Carbolic acid, 5
   Distilled water, 100

   B. Methylene blue, 2
   Sulphuric acid, 25
   Distilled water, 100

The prepared cover glass remains 10 minutes in solution A, is washed in water, dried, and is placed for 5 minutes in solution B. It is again washed in water and dried. The specimen should now have a light blue color. If it is still red in parts, it should be put into solution B for from 1 to 3 minutes, washed, and dried. It is then mounted in oil of cloves or Canada balsam and examined. The tubercle bacilli are red; everything else in the field is blue. This method is to be recommended for its accuracy and its clear results.

The Bacteria of Diagnostic Importance

The bacteria of pus. — *a. Staphylococcus;* is arranged in irregular masses; is stained by all the aniline dyes. When a pure culture shows a growth of yellow colonies, it is known as *Staphylococcus pyogenes aureus;* when white colonies are formed in pure culture, the germ is given the name of *Staphylococcus pyogenes albus.* It may appear in any suppurative process, abscesses, phlegmona, purulent inflammations of the serous membranes, otitis, osteomyelitis, purulent inflammatory conditions following typhoid fever, etc.

b. Streptococcus; is arranged in chains and occurs in many purulent processes. The inflammatory conditions provoked by the streptococcus are more intense than those of the staphylococcus, and have a tendency to make deeper inroads upon the system.

Streptococci are the etiological elements of erysipelas; they, as well as the staphylococci, produce the various forms of sepsis, particularly *puerperal sepsis*. They also cause endocarditis, croupous pneumonia, etc.

At present, the different varieties of streptococci are regarded as morphologically identical, differing only in the degree of virulence they manifest. The virulence of a species is determined by animal experimentation.

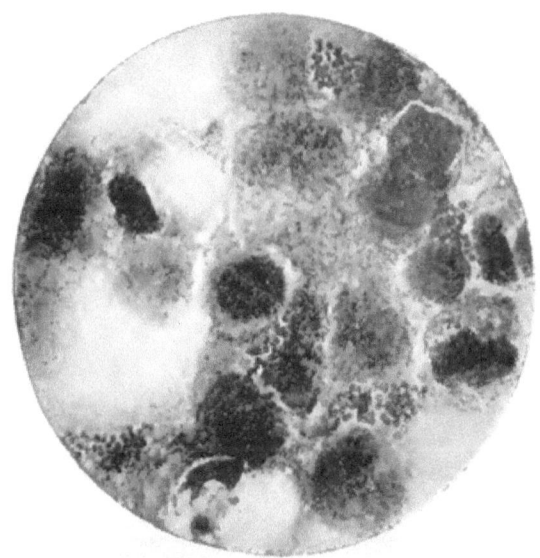

FIG. 54. — STAPHYLOCOCCI.

Gonococci (*Neisser*) (Fig. 55) are diplococci arranged like pairs of coffee-beans. They often completely fill the protoplasm of pus-cells, leaving only the nucleus free. Apparently, they are to be found only in the pus of gonorrhœa or of gonorrhœal infection (gonorrhœal conjunctivitis, cystitis, gonitis, pleurisy, and endocarditis). The presence of gonococci is accepted by most specialists as a proof of the correctness of diagnosis in doubtful cases of urethritis and leucorrhœa.

Meningococci (*Diplococci intracellulares*) (*Weichslebaum-Jäger*) are diplococci shaped like a roll, chiefly lying within

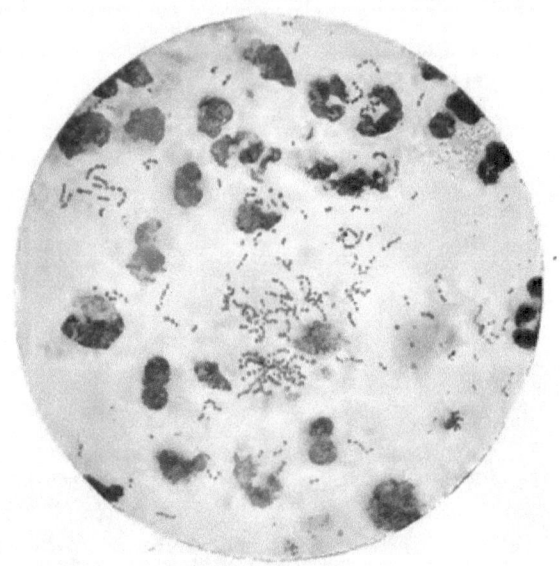

Fig. 54. — Streptococci.

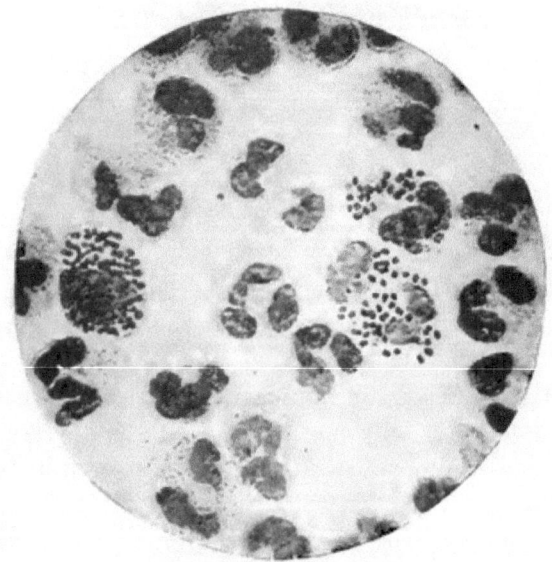

Fig. 55. — Gonococci.

the cells, resembling gonococci, abundantly demonstrable during life in cases of cerebro-spinal meningitis by means of lumbar puncture.

Pneumococci, the diplococcus of pneumonia (A. Fränkel) (Fig. 56), are lancet-shaped diplococci which are regularly found in the fibrinous exudate of lungs affected with pneumonia and in the sputum accompanying pneumonia. The microscopic examination usually suffices to recognize the pneumococcus with certainty; the diagnosis is rendered more certain, however, by making a culture, and by the inoculation of the germ into rabbits which die with a typical septicæmia. The absence of the pneumococcus in the sputum argues against a diagnosis of pneumonia; its presence does not make the diagnosis absolutely certain, since it is also found in the sputum of healthy individuals. The presence of the pneumococcus in the pus of empyema makes the etiology of this disease dependent upon a former pneumonia. The pneumococcus may be responsible for purulent processes in other parts of the body (meningitis, otitis, salpingitis, etc.).

Typhoid bacilli (*Eberth*) (Fig. 57) are short rods with curved ends. They are found in patients suffering from typhoid fever in the characteristic intestinal ulcers, in the mesenteric glands, in the spleen, and, in severe cases, they may be found in other organs as well as in abscesses developing late in the disease. From the time of the beginning of the separation of the crust of the intestinal ulcer, typhoid bacilli may be found in the stools. The ordinary cover-glass preparation is not characteristic, however, and even the usual culture methods are unsatisfactory, because the bacterium coli is almost identical in culture with the typhoid bacillus.

Typhoid bacilli and the bacterium coli grow upon the culture medium — suggested by *Elsner* — of *potassium iodide*-potato-gelatine, in differential forms, while the other bacteria of the fæces do not develop thereon. By *Elsner's* method, the diagnosis of typhoid

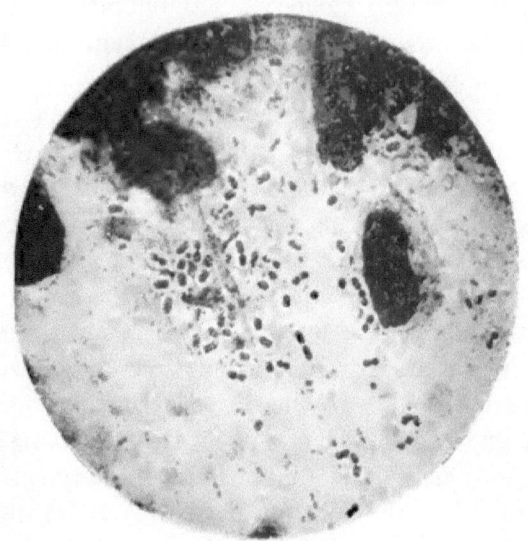

FIG. 56. — PNEUMOCOCCI.

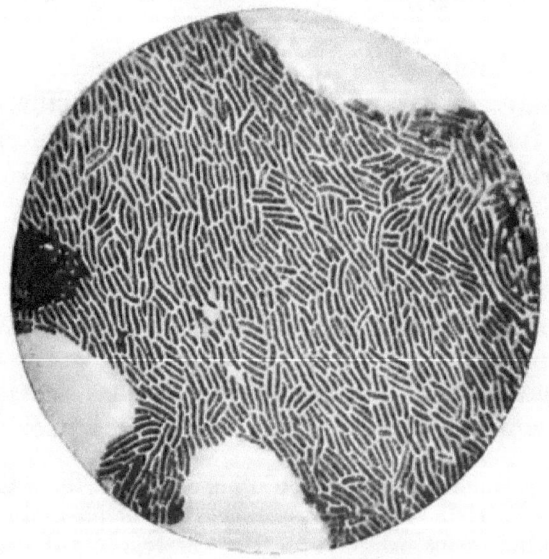

FIG. 57. — PURE CULTURE OF TYPHOID BACILLI.

fever may often be established by the bacteriological examination of the fæces in from 24 to 48 hours.

Very recently it has been shown that the blood-serum of animals which have been immunized against typhoid bacilli possesses the specific property of affecting typhoid bacilli in such a manner that their motion is inhibited, that they resolve themselves into clumps, and become finally disintegrated (agglutination, *Pfeiffer, Gruber*). The blood-serum of persons suffering from typhoid fever has the same property (*Widal*). *Widal's reaction* for the diagnosis of typhoid fever consists in the fact that when a drop of blood-serum taken from a typhoid patient is mixed with a small quantity of a bouillion culture of the typhoid bacillus, placed in a hanging drop under the microscope, and then observed, agglutination occurs. The same reaction may be obtained when a drop of such blood is added to a 24-hour old bouillion culture of this bacillus in a test-tube, and the same placed for 12 hours in a thermostat. It is then examined for agglutination. However, the proportion of serum to bouillion culture must always be less than 1 to 25. Up to the present time, it would appear as though a positive reaction occurs only in typhoid fever.

Bacterium coli commune, short rods, scarcely to be distinguished from typhoid bacilli on a cover-glass preparation or in the ordinary culture, is found in abundance in the contents of the large intestine of man. It may produce any inflammatory or purulent condition which appears in the neighborhood of the intestines or the genito-urinary tract (peritonitis, appendicitis, abscess of the liver, cystitis, pyelitis), and may even lead to a general sepsis.

Cholera bacilli (*Koch*) (Fig. 58) are short, curved rods (comma bacilli, vibriones). They are found in great quantities in the stools of cholera patients. A diagnosis, however, cannot be made from these bacilli except in culture; for some similarly shaped saprophytes may also appear in the fæces.

The diagnosis of Asiatic cholera can be made certain only by the demonstration of the presence of the specific microorganism. With a platinum needle, a whitish particle of mucus is selected from the suspected stool and is shaken in a test-tube with melted gelatine; from this test-tube a second one, also containing melted gelatine, is inoculated with a platinum loop. Both test-tubes are emptied into Petri dishes. Upon the hardening of the gelatine, these dishes are kept at a temperature of 22° C. in the thermostat. In from 24 to 48 hours the colonies of cholera bacilli are recognizable. They liquefy the gelatine, and, by so doing, form

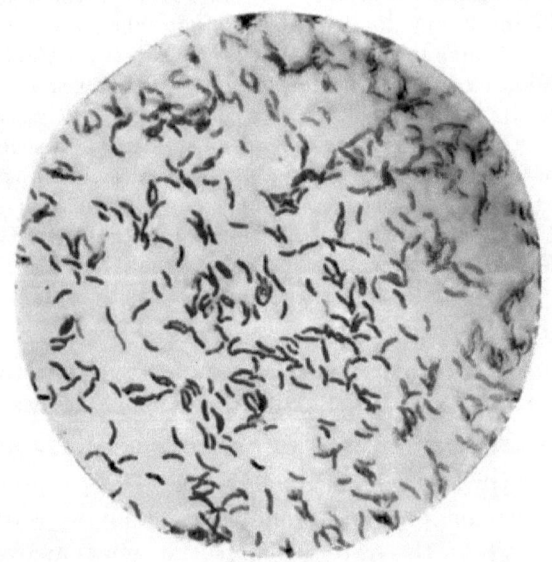

FIG. 58. — PURE CULTURE OF CHOLERA BACILLI.

characteristic, funnel-shaped depressions. With a low magnifying power, the individual colonies may be distinguished by their weak lustre, by their arrangement, which bears a resemblance to fragments of broken glass, and by their irregular edges. The depressions in the gelatine are more clearly seen when the micrometer screw is employed.

If there are but few vibrios microscopically demonstrable in the fæces, the plate-culture method of *Schottelius-Koch* must be employed. A small quantity of fæces is placed in a nutritive culture medium of 1 per cent. peptone and ½ per cent. sodium

T

274   ANIMAL AND VEGETABLE PARASITES   CHAP.

chloride. In the thermostat these vibrios increase much more extensively in this medium than the other fæcal bacteria. As a result of their need for oxygen, they accumulate on the surface, where they form a film in about 12 hours. A small particle of this film is then inoculated on a gelatine plate.

In addition to the culture method, experiments on animals assist in the diagnosis. The *Pfeiffer* reaction is differentially diagnostic: cholera bacilli rapidly die in the peritoneal cavity of highly immunized guinea-pigs. The smallest quantity of the blood-serum of highly immunized animals will agglutinate cholera bacilli (*Gruber*).

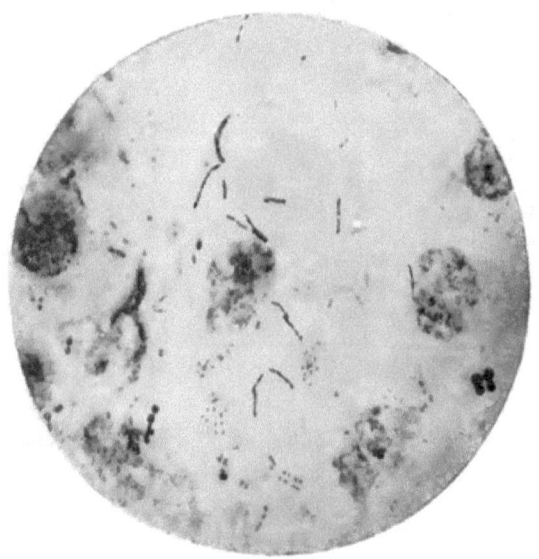

FIG. 59. — BACILLI IN TUBERCULAR SPUTUM.

Tubercle bacilli (*Koch*) (Fig. 59) are narrow rods, about three-fourths of the size of a red blood-cell. Their staining reactions are characteristic (see p. 266). The presence of tubercle bacilli in an organ is an absolutely certain proof of the existence of tuberculosis there. They are found in the sputum (pulmonary tuberculosis), in the urine (tuberculosis of the genito-urinary tract), in the blood (miliary tuberculosis), in the fæces (tuberculosis of the intestine), in pus (tuberculosis of the bones, empyema, etc.), and in the skin (lupus).

Tubercle bacilli in the *stools* of tubercular patients may be derived from sputum which has been swallowed, and do not indicate necessarily tuberculosis of the intestine.

In the smegma of the prepuce and labia, are many short rods which give some of the characteristic staining reactions of tubercle bacilli (*Smegma bacilli*). These must be borne in mind when there is a suspicion of urogenital tuberculosis. They lose their color in absolute alcohol in 1 minute, while tubercle bacilli retain their stains for several minutes.

The spirilla of relapsing fever (*Obermeier*) (Fig. 60) are found in the blood of patients suffering with relapsing fever. They are seen only when fever is present, and with a high magnifying power they may be seen in active movement when unstained. Cover-glass preparations may be stained with any of the aniline dyes.

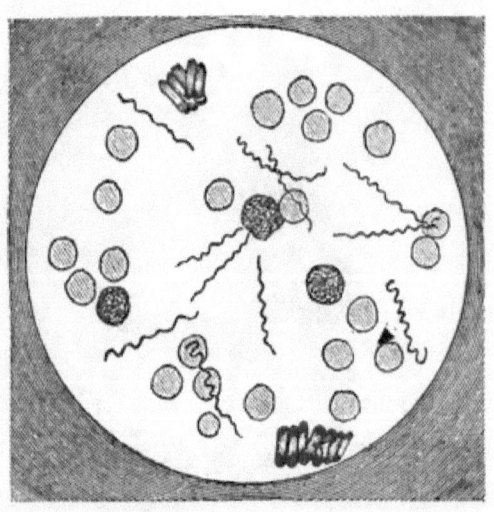

FIG. 60.—SPIRILLA FROM A CASE OF RELAPSING FEVER DURING THE FEVER.

Anthrax bacilli (*Davaine*) are large, thick rods, quite easily recognized in a cover-glass preparation. The diagnosis is rendered certain by inoculation upon a mouse, which dies in from 1 to 2 days after the inoculation. The blood of the mouse is found filled to overflowing with anthrax bacilli. Diagnostically, the anthrax bacillus is important because in man it is productive of large carbuncles, the specific nature of which can be recognized only by the demonstration of the bacilli.

276 ANIMAL AND VEGETABLE PARASITES CHAP.

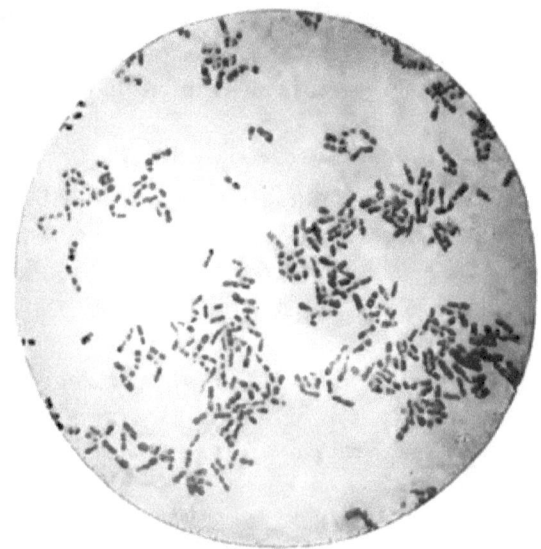

FIG. 61.—DIPHTHERIA BACILLI.

Diphtheria bacilli (*Löffler*) (Fig. 61) are short, narrow rods, the etiological factors of diphtheritic inflammation, in which a deep necrosis of the mucous membrane goes hand in hand with the formation of a membrane. If the bacilli are not very virulent, they may call forth a simple catarrhal or fibrinous inflammation of the mucous membrane. A pure culture of diphtheria bacilli may sometimes be obtained by passing a small piece of diphtheritic membrane over several culture plates containing solidified blood-serum. In this culture medium, a shining surface develops in two days, which on microscopic examination of stained cover-glass preparations shows a collection of short rods.

In all doubtful cases of follicular angina, the membrane must be examined microscopically and by means of cultures for diphtheria bacilli. The presence of bacilli in the membranes settles the diagnosis.

Glanders bacilli (*Löffler*) may be found in profusion in the nodes of glanders; they grow on potatoes in characteristic brown patches. If the pus of glanders or a pure culture of the bacilli is injected into the abdominal cavity of a male guinea-pig, its testicles undergo purulent inflammation (*Strauss*).

Influenza bacilli (*Pfeiffer*), very minute rods, are contained in great numbers in the bronchial mucous membrane and in pneumonic foci in influenza. Agar is a good culture medium when previously covered with fresh blood. The colonies develop as very minute drops, as clear as water and scarcely appreciable with the naked eye.

Tetanus bacilli (*Nicolaier-Kitasato*) are demonstrable in the wounds or purulent inflammations which usually evoke tetanus. The demonstration of tetanus bacilli is chiefly of theoretical interest because the clinical diagnosis is always easily established.

Actinomyces. — The ray fungus is of diagnostic importance, since it is the etiological factor of *actinomycosis*, an infectious disease running its course with chronic suppuration. The ray fungus forms, in dense masses, yellowish granules of the size of a poppy. On microscopic examination, they separate into chains of fungi with peripheral, club-shaped enlargements, resembling a bunch of grapes in their arrangement.

CHAPTER XIII

THE RÖNTGEN RAYS AS DIAGNOSTIC AIDS

In 1895 Professor Röntgen of Würzburg discovered that, by the discharge of induced electric currents through a vacuum tube (*Crookes's*), peculiar rays of light emanated from the cathode. These X-rays, so called by Röntgen, possess the remarkable property of penetrating solid objects, and are not subject to reflection or to refraction. Looked at through a screen covered with bario-platinum cyanide, in a dark room, the X-rays evoke a bright light; upon a sensitive photographic plate they produce impressions which may be developed by ordinary daylight.

The more solid the body, the greater resistance does it offer to the passage of the Röntgen rays. Wood is more easily permeable than metal, and paper than wood. The muscles of the human body allow less easy passage of the rays than the soft parts, and the bones are least permeable.

By placing the human body between a Crookes's tube and the screen covered by the bario-platinum cyanide, the more solid parts of the body appear upon the plate as dark shadows, while the soft parts are more or less light, depending upon their respective densities. The plate thus shows a shadow-picture of the interior of the body (*actinogram*), the bony apparatus appearing very distinctly, while the organs of the chest and abdomen are easily recognized in outline.

If the body is placed between the Crookes's tube and a photographic plate enclosed in a box, the soft parts appear dark in the negative, the denser parts light. The positive made from this shows, correspondingly, the bones, etc., dark, the soft parts more or less light.

Immediately upon the announcement of the discovery of the Röntgen rays, they began to be employed for medical purposes, and although scarcely a year and a half has elapsed since the publication of this phenomenon, it may be said that by its aid many internal diseases can be diagnosticated early or, at least, that a diagnosis can be placed beyond doubt. As the apparatus is comparatively easy to acquire and as the technic of its use is not difficult, many physicians are experimenting with the X-rays, and further advances may be awaited with certainty.

The heart. — Dilatation and pericarditis may be diagnosticated, as the size of the heart can be seen with the aid of the rays. The contractions of the heart can be distinctly witnessed upon the fluorescent screen, and this enables us to form a judgment as to the strength and rhythm of the cardiac beat which can be substantiated by the examination of the apex-beat and the pulse.

The blood-vessels. — With the aid of the Röntgen rays, calcification of the arteries can be recognized. The picture of this phenomenon on the cadaver is very striking, but it is not so sharply marked in the living subject. A more practical use can be made of the fluoroscope by observing dilatation of the arteries, by which means the diagnosis of *aneurysm of the aorta* may be made much earlier than has hitherto been the case. Aneurysm of the aorta has been seen, for example, before any local symptoms have appeared. A differential diagnosis from other mediastinal tumors can be made by the visible pulsation.

The lungs. — The lungs are visible as filling the thoracic cavity as a weak shadow, and their size can thus be easily judged. The diagnosis of emphysema can thus be made without further difficulty. Pleuritic effusions, adhesions, and large infiltrations are readily recognized by the deeper shadows they produce. It is not yet certain if small infiltrations whose presence we have as yet no objective methods of demonstrating, can be made recognizable by means of the skiagraph.

Tumors of the lungs or of the pleuræ can be diagnosticated with certainty much earlier than previously. This is, beyond doubt, an important step in the progress of diagnosis.

The abdominal organs. — The observations so far upon the abdominal organs have not been very satisfactory. Large tumors have been easily seen, but an early diagnosis of them has not yet been accomplished. It is frequently impossible to differentiate the outlines of the various organs. The *œsophagus* and *stomach* can be more readily distinguished by the preliminary passing of a stomach tube surrounded by a spiral of metal or filled with small shot. Even this measure has not furthered diagnosis. It is not yet established with certainty to what extent *renal, vesical,* or *biliary calculi* can be demonstrated. The greatest progress has been made in the use of this new photography in the diagnosis of diseases of the **bones** and **joints**, and in demonstrating the presence of foreign bodies. In these instances, surgery has been a greater gainer than medicine. For internal medicine, the presence of *caries of the vertebræ,* the bony changes of *rhachitis,* the anchylosed joints and the deformed epiphyses of chronic articular rheumatism, are of importance. It would seem that the Röntgen rays will offer a means of making a differential diagnosis between *gout* and *arthritis deformans,* so frequently difficult to determine.

INDEX

ABDOMEN, 20, 95; diseases of. 99;
  distention of, 95; tumors of, 98.
Abducens paralysis, 47.
Acarus folliculorum, 262; scabiei,
  261.
Accessorius paralysis, 47.
Acetone and aceto-acetic acid, 199.
Achilles tendon reflex, 55.
Achorion Schoenleinii, 262.
Acid, hydrochloric test for, 86; hippuric, 208; lactic, test for, 86; succinic, 258; sulphanilic, 201.
Acoustic nerve, lesion of, 45.
Actinography, 278.
Actinomyces, 150, 277.
Adenoid vegetations, 108.
Ægophony, 141.
Agony, 15.
Agraphia, 50.
Air, complementary, 127; reserve, 127; residual, 127; tidal, 127.
Albumin, qualitative tests of, 186; quantitative test of, 187; albuminometer, 187.
Albuminuria, 185; cardiac, 161; cyclic, 186; intermittent, 186; physiological, 185; spurious, 185.
Albumosuria, 188.
Alcoholism, 44, 50, 52.
Alexia, 50.
Alimentary glycosuria, 193.
Alloxan bodies, 208.
Amaurosis, 46.
Ambylopia, 46.
Ammonium urate, 213.
Amœba coli, 254.
Amphoric resonance, 134; breathing, 138.
Amygdallitis, follicular, 29.
Amyotrophic lateral sclerosis, 68.

Anacidity, 76.
Anæmia 250; pernicious, 251; secondary, 252.
Anæsthesia, 57; of pharynx, 109.
Analgesia, 58.
Anamnesis, 4.
Anarthria, 49.
Anatomy of brain and spinal cord, 38 seq.
Anchylostoma duodenale, 259.
Angina follicularis, 29; pectoris, 160.
Anguillula intestinalis, 261.
Angulus Ludovici, 123.
Ankle clonus, 55.
Anode, 60.
Anthrax, bacillus of, 275.
Aorta, aneurysm of, 180; insufficiency of, 178; stenosis of, 178.
Aortitis, 180.
Apathy, 35.
Apex-beat, 162; displacements of, 162; systolic retraction of, 164.
Aphasia, 49.
Aphonia, 109, 118.
Apnœa, 9.
Apoplexy, 43.
Appendicitis, 98.
Appetite, 76.
Arsenic, test for, 217.
Arterial sounds, 170.
Arterio-sclerosis, 179.
Arthrozoa, 254, 260.
Articular sense, 59.
Ascaris lumbricoides, 259.
Ascites, 96.
Aspergilli, 150, 263.
Asthma, bronchial, 125; cardiac, 161; crystals of, 148; dyspeptic, 126; nasal, 126; uræmic, 126.
Ataxia, 50.

282 INDEX

Atelectasis, 132.
Athetoid movements, 53.
Atrophy in paralysis, 42, 68.
Auscultation of heart, 167; thorax, 136; vessels, 170; voice, 141.
Auscultatory percussion, 134.
Autoscopy, 113.
Axillary lines, 128.

BACILLI, 264.
Bacteria, 264.
Bacterium coli, 272.
Basedow's disease, 172, 240.
Biedert's method, 151.
Biermer's change of sound, 135.
Biliary colic, 104; ducts, catarrh of, 103.
Biuret reaction, 87.
Bladder epithelium, 214; calculi in, 226; cancer of, 226; tumors of, 226.
Blood, 241; cells, red, 243; cells, red, nucleated, 247; cells, measurements of, 245; cells, staining of, 246; cells, white, 244; corpuscles, counting, 244; diseases of, 241, 250; examination of, 242; expectoration of, 142; hæmoglobin in, 248; in the stool, 92; in the urine, 189; plasmodium malariæ in, 27; -poisoning, 31; reaction of, 249; specific gravity of, 249; spectroscopy of, 249; tests for, 189; -vessels, diseases of, 179; vomiting of, 79.
Blue sputum, 145.
Boas's test meal, 86.
Böttcher's test for sugar, 194.
Bothriocephalus latus, 256.
Brachycardia, 171.
Brain, abscess of, 67; syphilis of, 67; tumors of, 67.
Breast, funnel-shaped, 123; pigeon's, 123; wedge-shaped, 123.
Bright's disease, 220.
Bronchial casts, 144.
Bronchiectasis, 152.
Bronchitis, 151; fetid, 157.
Bronchophony, 141.
Broncho-pneumonia, 153.
Bruit de pot fêlé, 134.

Bulbar paralysis, 68.
Bulimia, 76.
Burdach, columns of, 41.

CALCULI, biliary, 104; biliary, analysis of, 164; cystin, 227; renal, 225; vesical, 226; vesical, analysis of, 226; xanthin, 227.
Calomel stool, 92.
Caloric necessity, 230.
Caloric value of food-stuffs, 231.
Calves, cramp of, 52.
Capillary pulse, 178.
Caput Medusæ, 97.
Carbonate of lime, 212.
Carbonic oxide blood, 250.
Carcinoma of intestines, 100; liver, 104; œsophagus, 74; stomach, 90.
Cardiac area, bulging of, 163; asthma, 161; impulse, 163; murmurs, 167; œdema of, 161; thrills, 164.
Carotids, sounds in, 170.
Casts, bloody, 214; bronchial, 144; brown, 214; epithelial, 214; granular, 214; hyaline, 214; waxy, 214.
Catalepsy, 53.
Catarrh of biliary ducts, 103; intestines, 99.
Cathode, 60.
Cavities, 134; respiration in, 135; symptoms of, 135.
Cell-breathing, 136.
Cercomonas intestinalis, 253.
Cerebral abscess, 67; syphilis, 67.
Cestodia, 254.
Charcot-Leyden crystals, 148.
Chest, barrel-shaped, 123; circumference of, 127; pain in, 121.
Cheyne-Stokes breathing, 10.
Chills, 17.
Chlorosis, 250.
Choked disc, 45.
Cholera, bacilli of, 96, 272; stools of, 94.
Choreic movements, 53.
Chyluria, 202.
Circulatory apparatus, diseases of, 160.

INDEX 283

Cirrhosis of liver, 105.
Cocci, 264.
Colic, biliary, 104; mucous, 99; renal, 224.
Collapse, 14; temperature in, 14.
Color of the fæces, 91; skin, 8: sputum, 144; urine, 182.
Columns of Burdach, 41: Goll, 41.
Coma, 35.
Comma bacillus, 272.
Complementary air, 127.
Complexion, 7.
Concretio pericardii, 179.
Conjunctival reflex, 54.
Constipation, 93.
Convulsions, 50; clonic, 50: localized, 52; tonic, 50.
Corset-liver, 103.
Costal breathing, 124.
Costo-abdominal breathing, 124.
Cough, 111, 121; varieties of, 122.
Cover-glass specimens, 265; staining of, 265.
Cracked-pot sound, 134.
Cramp of leg, 52; sterno-cleido-mastoid, 52.
Cranial nerves, 45.
Cremaster reflex, 53.
Crisis, 18.
Crises, gastric, 69.
Croupous pneumonia, 23, 150.
Cryptogenic sepsis, 109.
Crystals of acid phosphate of calcium, 212; Charcot-Leyden, 148; of fatty acids, 148: oxalate of calcium, 211.
Curschmann's spirals, 148.
Curve of Ellis, 154; Damoiseau, 154.
Cyanosis, 8.
Cyclic albuminuria, 186.
Cysticercus cellulosæ, 255.
Cystin calculi, 227.
Cystitis, 225.

DEATH, signs of, 15.
Defervescence, 24.
Degeneration, reaction of, 64; complete, 66; partial, 66.
Delirium, 37.

Dextrose, 192.
Diabetes mellitus, 196, 238, 239; metabolism in, 238.
Diagnostic puncture, 154.
Diameter, costal, 123; sterno-vertebral, 123.
Diaphragm, paralysis of, 48.
Diarrhœa, 91, 93.
Diastase, 72.
Diazo-reaction, 200.
Dicrotism, 174.
Dilatation of heart, 165, 167.
Diphtheria, 30; bacillus of, 276.
Diphthongia, 110.
Diplococci, 264.
Diplopia, 46.
Distoma hæmatobium, 258; hepaticum, 258: lanceolatum, 259.
Dittrich's plugs, 149.
Diverticulum of œsophagus, 73.
Dorsal position, 5.
Double sensation, 58; sounds, 171.
Dropsy, 10; cachectic, 11; with albuminuria, 10; with cyanosis and dyspnœa, 10.
Drug eruptions, 12.
Dry râles, 139.
Dulness over lungs, 132; sternum, 167.
Dynamic sense, 59.
Dysentery, 94.
Dyspepsia, 76, 82, 90; nervous, 91.
Dysphagia, 111.
Dysphonia, 109.
Dyspnœa, 9; expiratory, 125.

ECHINOCOCCUS, 257; cysts, 257: of liver, 105; in sputum, 150.
Eclampsia, 51.
Ehrlich's stain for bacilli, 266; diazo-reaction, 200.
Einhorn's saccharimeter, 198.
Elastic fibres, 147.
Electric irritability, 59.
Electricity, cutaneous, sensibility to, 59.
Embryocardia, 172.
Emphysema, 152.
Empyema, 154; meta-pneumonic, 155.

Endocarditis, acute, 32; ulcerative, 32.
Enteritis membranacea, 100.
Enterogenous albumosuria, 188.
Enteroptosis, 84.
Eosinopile cells in blood, 248; sputum, 145.
Epidemic meningitis, 31.
Epilepsy, 50.
Epistaxis, 107.
Epithelium, alveolar, 145.
Eructations, 76.
Eruptions, 12.
Erysipelas, 22; cocci of, 267.
Esbach's albuminometer, 187.
Exacerbation, 16.
Exanthemata, 11, 19.
Exchange of food-stuffs, 232.
Expectoration, 111.
Expression, 6.
Extra-pericardial friction sounds, 170.
Exudation, hæmorrhagic, 155; purulent, 155; serous, 155.

FACE, appearance of, 7; color of, 7; expression of, 7.
Facial paralysis, 47.
Facies composita, 6; decomposita, 6.
Fæces, 91; amount of, 91; estimation of, 91; fat in, 92; nitrogen in, 236; vomiting of, 80.
Falsetto voice, 109.
Faradic current, 59.
Fastigium, 18.
Fatty heart, 177; kidney, 222.
Favus, fungus of, 262.
Features, 6.
Fehling's test for sugar, 194, 195.
Femoral artery, double sound in, 171.
Fermentation test for sugar, 194.
Ferric chloride reaction, 200.
Fever, character of, 19; continuous, 18; course of, 18; intermittent, 18, 27; inverted type, 17; malarial, 27; recurrent, 25; relapsing, 25; remittent, 18; scarlet, 21; stupida, 19; symptoms of, 18; type of, 18;
typhoid, 24; typhus, 25; versatilis, 19; of suppuration, 31.
Fibrin, 143.
Filaria sanguinis, 261; medinensis, 261.
Flea, 261.
Fleischl's hæmometer, 248.
Flukes, 253, 258.
Flushing, 7.
Food, metabolic changes in, 230; nutritive value of, 234.
Forced movements, 53.
Fränkel-Gabbet's staining method, 267.
Friction sounds, extra-pericardial, 170; pericarditis, 170; pleural, 140.
Fuliginous lips, 20, 70.
Functional paralysis, 42.
Fungi, 262.
Funnel chest, 123.

GABBET's staining method, 267.
Gagging, 110.
Gait, 50.
Gall-stones, 104; cholesterin in, 104.
Galvanic current, examination by, 60.
Gangrene of lungs, 158.
Garrod's thread test, 207.
Gas, escape of, into abdominal cavity, 98.
Gastric crises, 69, 78.
Gastritis, acute, 90; chronic, 91.
Gastrodiaphany, 84.
Gerhardt's ferric chloride reaction, 200; change of sound, 135.
Gibbus, 123.
Gigantocytes, 247.
Glanders, bacillus of, 276.
Glenard's disease, 84.
Glossitis, 71.
Glossopharyngeal nerve, paralysis of, 47.
Glottis, dilatation of, 113; spasm of, 116; stenosis of, 111.
Glycosuria, transitory, 195.
Gmelin's test, 104, 191.
Goll's columns, 41.
Gonococci, 268.
Gout, 240.

Gram's stain, 266.
Gubler's hemiplegia, 44.
Guiac test for blood, 189.
Gums in lead-poisoning 72.

HABITUS, 5; apoplecticus, 5; neurasthenicus, 5; paralyticus, 125; phthisicus, 5.
Hæmatemesis, 79.
Hæmatoidin crystals, 148.
Hæmaturia, angioneurotic, 190; of bleeders, 190.
Hæmoglobin, estimation of, 248.
Hæmoglobinuria, 189, 191.
Hæmoptysis, 142.
Hæmorrhagic exudation, 155.
Hæmosiderin reaction, 147.
Harrison's furrow, 127.
Häser's coefficient, 182.
Hay fever, 126.
Headache, 107; diffuse, 37; frontal, 107.
Heartburn, 76.
Heart, auscultation of, 167; dilatation of, 165, 167; diseases of, 176; disease, urine in, 176; disease-cells, 147; displacement of, 166; disturbances of, neurotic, 161; dulness over, 165; hypertrophy of, 166; inspection of, 162; murmurs in, 168; percussion of, 165; sounds, 167; sounds accentuated, 168.
Heller's blood test, 189; test for albumin, 187.
Hemianopsia, 45.
Hemianopic iridoplegia, 46.
Hemiplegia, 42; embolic, 43; apoplectic, 43; toxic, 43.
Hemiopia, 45.
Hemiopic reaction of pupil, 46.
Hepatization, 132, 153.
Hepatogenous albumosuria, 188.
Herpes, 19.
Hippuric acid, 208.
Hoarseness, 109.
Hodgkin's disease, 251.
Hofmeister's test for peptone, 188.
Hydrobilirubin, test for, 192.
Hydrochloric acid, tests for, 86.

Hydronephrosis, 225.
Hyperacidity, 76, 84.
Hyperæsthesia, 57; of pharynx, 109.
Hyperleucocytosis, 250.
Hyperpyretic temperature, 17.
Hypertrophic cirrhosis of liver, 105.
Hypochondrium, 112.
Hypoglossus nerve, paralysis of, 47.
Hypostatic congestion, 132.
Hypoxanthin, 208.

ICTERUS, 7, 101; with polycholia, 8; without polycholia, 8; gravis, 8, 101; simplex, 7, 101.
Idiopathic heart-disease, 176.
Ileus, 95, 100.
Imperative movements, 53.
Incubation, 19.
Indican, 191, 208.
Indigo red, test for, 209.
Infarct of lung, hæmorrhagic, 158; kidney, hæmorrhagic, 224; spleen, hæmorrhagic, 106.
Infectious diseases, pathognomonic symptoms of, 19.
Infiltration, 132.
Influenza, 29; bacillus of, 277.
Infusoria, 253.
Insanity, 37.
Inspection of heart, 162; stomach, 81; thorax, 122.
Intense breathing of Kussmaul, 36.
Intention spasm, 52.
Intermittent fever, 18.
Intestines, carcinoma of, 100; catarrh of, 99; diseases of, 91; obstruction of, 95; trichinæ in, 260.

JACTITATION, 6.
Jaundice, 7, 101.
Jendrassik's trick, 55.
Joint sense, 58.

KIDNEY, amyloid, 223; atrophic, 223; diseases of, 220; epithelium of, 214; fatty, 222; hæmorrhage from, 190; infarct in, hæmorrhagic, 224; passive congestion of, 223; percussion of, 224; wandering, 225.
Kjeldahl's test for urea, 206.

Kreatinin, 208.
Kühne's peptone, 188.
Kypho-scoliosis, 123.
Kyphosis, 123.

LABZYMOGEN, 89.
Lactic acid, test for, 86.
Larynx, muscles of, 113; nerves of, 113; pain in, 111; stenosis of, 111; syphilis of, 115; tuberculosis of, 115; tumors of, 115.
Laryngitis, acute, 114; chronic, 114.
Laryngoscopic examination, 112.
Legal's test for acetone, 200.
Leptothrix in sputum, 147.
Leucæmia, lymphatic, 251; myelogenous, 251; splenic, 251.
Leucin, 212.
Leucocytes, 244; mononuclear, 247; polynuclear, 248; in urine, 213.
Leucocytosis, 250.
Leyden's crystals, 148.
Lieben's test for acetone, 200.
Liebig's test for urea, 205.
Lips, the, 70.
Liver, abscess of, 104; amyloid, 105; acute yellow atrophy of, 104; carcinoma of, 104; chronic congestion of, 105; cirrhosis of, atrophic, 105; cirrhosis of, hypertrophic, 105; dulness, 102, 103; echinococcus of, 105; movable, 103; palpation of, 102; passive congestion of, 105; percussion of, 102; venous pulse in, 164, 174.
Livor, 9.
Localization, sense of, 59.
Locomotor ataxia, 69.
Lordosis, 123.
Lungs, abscess of, 159; anthracotic, 145; echinococcus of, 159; gangrene of, 158; infarct of, hæmorrhagic, 158; inflammation of, 153; percutory limits of, 129; syphilis of, 159; topography of, 128; tuberculosis of, 156; tumor of, 159; tympanitic percussion note over, 133; vital capacity of, 126.
Lymphocytes, 247.
Lysis, 18.

MACROCYTES, 243.
Malarial fever, 27; neuralgia, 29.
Mamillary line, 128.
Maximum thermometer, 17.
Measles, 21.
Measurement of blood corpuscles, 245.
Median nerve, paralysis of, 48.
Mediate percussion, 134.
Megalogastria, 84.
Melancholia attonita, 53.
Membranous enteritis, 100.
Menière's disease, 37.
Meningitis, cerebro-spinal, 30; epidemic, 31; tubercular, 31, 67.
Meningococci, 268.
Mensuration, 127.
Metabolic balance-sheet, 237, 239.
Metabolism, anomalies of, 233; disorders of, 229; equilibrium of, 232; nitrogenous, 232; normal, 229.
Metallic note, 131.
Metamorphic respiration, 138.
Metapneumonic empyema, 155.
Meteorism, 95.
Methæmoglobin, 250.
Methyl reaction, 86.
Microcytes, 243.
Micrococci, 264.
Micro-organisms, 264; in urine, 216.
Microsporon furfur, 263.
Migraine, 46.
Miliary tuberculosis, 30.
Milk, test meal of, 85.
Minute thermometer, 17.
Miserere, 80, 95.
Mitral insufficiency, 178; stenosis, 178.
Mohrenheim's space, 127.
Monoplegia, 42.
Moore's test (quantitative for sugar), 196.
Morbilli, 21.
Morbus Basedowii, 240; Brightii, 220; maculosus, 11.
Morning vomiting, 80.
Motor irritation, 51; paths, 38.
Moulds, 262.
Mouth, 71; breathing, 108; epithelium of, 145.

Movable liver, 103; spleen, 106.
Movements forced, 53.
Mucorinæ, 263.
Mucous membrane, reflexes of, 54.
Multiple sclerosis, 69.
Mumps, 31.
Murexide test, 207.
Murmurs, accidental, 169; cardiac, 168; diastolic, 169; functional, 169; presystolic, 169; respiratory, 136; systolic, 169.
Muscle sense, 59.
Musical timbre (cardiac), 168.
Mydriasis, 47.
Myelitis, 68; cervicalis, 68; dorsalis, 68; lumbalis, 69.
Myelogenous leucæmia, 251.
Myosis, 47.
Myotonia congenita, 52.
Myotonic reaction, 52.
Myxœdema, 240.

NASAL voice, 110.
Neck, stiffness in, 18, 31.
Nematoidia, 259.
Nephritis, acute, 222; chronic, 222; suppurative, 224; urica, 211.
Nephrolithiasis, 225.
Nerves, cranial, 45.
Nervous dyspepsia, 91; system, 33; system, status of, 33.
Neuralgia, 57, 106; malarial, 29.
Neuritis, 44, 50.
Neuroses, reflex nasal, 108.
Nipple reflex, 54.
Nitrogen in fæces, 236.
Nitrogenous equilibrium, 232.
Nose-bleed, 107.
Nose, diseases of, 107.
Nutrition and strength, condition of, 4.
Nylander's test for sugar, 194.
Nystagmus, 53.

OCHRE, yellow sputum, 144.
Œdema, 12.
Œsophagoscopy, 73.
Œsophagus, carcinoma of, 74; diverticula of, 73; stricture, 72.
Œsophagitis, 74.
Oidium albicans, 72, 263.
Oil test (motility of stomach), 90.
Opisthotonus, 51.
Organic heart murmurs, 169.
Otitis media, 22.
Ova of animal parasites, 254 seq.
Oxalate of lime, 211.
Oxyuris vermicularis, 259.
Ozæna, 107.

PACHYDERMIA laryngis, 116.
Pain sense, 58.
Pallor, 7; eximius, 7.
Palpation of heart, 162; stomach, 81.
Palsy, lead, 72; nuclear, 47.
Paradoxical contraction, 56.
Paræsthesia, 57.
Paralysis, 41; abducens, 47; agitans, 53; bulbar, 68; cerebral, 67; of children, essential, 69; of cranial nerves, 45; crossed, 44; diaphragmatic, 48; Erb's, 47; facial, 47; functional, 41; glosso-pharyngeal, 47; hypoglossal, 47; intensity of, 49; median, 48; nuclear, 47; oculomotor, 46; olfactory, 45; optic, 45; peroneal, 48; phrenic, 48; pneumogastric, 47; posticus, 119; progressive bulbar, 68; radial, 48; recurrent, 46; recurrent laryngeal, 119; spinal, 47; spinal spastic, 68; tibial, 48; trigeminal, 47; trochlear, 47; ulnar, 48; vagus, 48; of vocal cords, 116.
Paralytic thorax, 122.
Paraplegias, 42, 44.
Parasites, 253.
Parasternal line, 128.
Paresis, 41.
Parotitis epidemica, 31.
Paroxysmal hæmoglobinuria, 191; tachycardia, 171.
Patellar reflex, 55.
Pathogenic micro-organisms, 267.
Pectoral fremitus, 140.
Pectoriloquy, 141.
Pectus carinatum, 123.

Pediculi, 261.
Peliosis, 11.
Penicillium glaucum, 262.
Pepsin, 88.
Peptone, test for, 188.
Peptonuria, 188.
Percussion of the heart, 165; kidney, 224; liver, 102; spleen, 106; stomach, 82; thorax, 129.
Pericardial friction sounds, 170.
Pericarditis, 179.
Perigastritis, 82.
Period of expulsion, 163.
Periomphalitis, 98.
Peritoneum, diseases of, 91.
Peritonitis, 96, 100; chronic, 97, 100; perforation, 99.
Perityphlitis, 98, 100.
Pernicious anæmia, 251.
Perturbatio critica, 18.
Pertussis, 122.
Petechiæ, 12.
Pflüger's test for urea, 205.
Pharynx, 109; hyperæsthesia of, 109.
Pharyngitis, 110.
Phenyl hydrazine test for sugar, 195.
Phosphate of calcium, 212.
Phosphates, triple, 212.
Phosphatic calculi, 226.
Phthisis pulmonum, 156.
Physiological albuminuria, 185; glycosuria, 195.
Pityriasis versicolor, 263.
Pleura, adhesions of, 140.
Pleuritis exsudativa, 154; retrahens, 155; sicca, 154.
Pneumococci, 270.
Pneumonia, 23, 153; congestive stage of, 23.
Pneumo-pericardium, 166.
Pneumo-thorax, 157.
Pock-marks, 26, 27.
Poikilocytes, 243.
Polarization, 198.
Poliomyelitis, 69.
Portal vein, thrombosis of, 97.
Position of patient, 6.
Posticus paralysis, 119.
Pregnancy, 77.

Pressure points, 57; sense, 58.
Presystolic murmurs, 169.
Prodromal stage, 19; vomiting, 77.
Proglottides, 254.
Progressive bulbar paralysis, 68; spinal atrophy of muscles, 42.
Protozoa, 253, 254.
Pseudo-crisis, 23.
Pseudo-leucæmia, 251.
Psychoses, 36, 37.
Puerperal albumosuria, 188.
Pulmonary artery, insufficiency of, 179.
Pulse, the, 13, 171; anacrotic, 174; arythmia of, 172; capillary, 178; curve, 174; dicrotic, 174; elastic elevation of, 174; excursion of, 175; frequency of, 19; frequency of, in fever, 18; hardness of, 173; over-dicrotic, 174; rapidity of, 171; rhythm of, 172; size of, 173; slowing of, 173; tension of, 173; velocity of, 173; venous, 164.
Pulse-wave, 174; anacrotic, 174; dicrotic, 174; katacrotic, 174; post-dicrotic, 174.
Pulsus alternans, 172; bigeminus, 172; celer et altus, 173; frequens, 171; paradoxus, 173 rarus, 171; trigeminus, 172.
Pupil, hemiopic reaction of, 46.
Purulent sputum, 142.
Pus, bacteria of, 264.
Pyæmia, 31.
Pyelonephritis, 224.
Pylorus, hypertrophy of, 82; stricture of, 82.
Pyogenic albumosuria, 188.
Pyramidal tract, 40.
Pyrosis, 76.
Pyuria, 213.

Rales, 139; crepitant, 139; dry, 139; metallic, 140; moist, 139; ringing, 139; sibilant, 139; sub-crepitant, 139.
Raspberry tongue, 20.
Reaction of blood, 249; urine, 183.
Rectus abdominis, cramp of, 51.
Recurrent fever, spirilla of, 275.

INDEX

Red blood-cells, 243; color of, 244; nucleated, 247; number of, 243.
Reflex, abdominal, 54; Achilles tendon, 55; cremaster, 54; foot clonus, 55; functions, 56; gluteal, 54; loss of, 55; mamillary, 54; patellar, 55; plantar, 54; skin, 54; tendon, 54.
Reflexes, 35, 53; abolition of, 55.
Relapsing fever, 25; spirilla of, 275.
Remittent fever, 18.
Renal hæmophilia, 190.
Reserve air, 127.
Residual air, 127.
Respiration, 136; amphoric, 138; bronchial, 137; broncho-vesicular, 138; cavernous, 137; costal type, 124; costo-abdominal type, 124; frequency of, 124; interrupted or jerky, 137; metamorphic, 138; puerile, 136; slowing of, 124; systolic, vesicular, 136; tracheal, 137; uncertain, 138; vesicular, 136; vesicular, sharpened, 136.
Respiratory air, 127; changes, 127; changes in sound, 135; movements, 124; murmur, 136.
Retropharyngeal abscess, 109.
Rheumatism, acute articular, 31.
Rhizopods, 253.
Rhythmical contractions, 52.
Risus sardonicus, 51.
Romberg's symptom, 50.
Röntgen rays, 278.
Roseola, 9.
Rosenbach's reaction for indigo red, 209.
Rouleaux formation, 243.
Round worms, 259.
Rubiginous sputum, 144.
Rubner's test for sugar, 194.

SACCHARIMETER, Einhorn's, 198.
Saccharomycetes, 262.
Saliva, 72.
Salkowski's test for peptone, 189.
Salol test of motility of stomach, 89.
Santonin in urine, 219.
Saprophytes, 264.

Sarcina pulmonum, 116.
Scapular line, 128; reflex, 54.
Scarlatina, 21.
Schizomycetes, 264.
Sclerosis, 69.
Scolex, 254.
Scoliosis, 123.
Scurvy, 12.
Secondary anæmia, 252.
Sedimentum lateritium, 210.
Semilunar space, 83.
Sensation, disturbances of, 57; test of, 58.
Sensorium, 19, 35.
Sensory paths, 40.
Sepsis, 31.
Septicæmia, 32.
Serous sputum, 142; exudation, 155.
Side position, 5; pain in, 121.
Signs of death, 13.
Singultus, 78.
Skin, bronze color of, 8; dryness of, 13; reflexes of, 54; temperature of, 12.
Small-pox, 26.
Smegma bacilli, 275.
Sneezing, 56.
Sodium urate, acid, 210.
Soor, 72.
Sound, changes in auscultatory, 130; qualities of, 129.
Spaces, complementary, 130.
Space sense, 58.
Spasm, 50; of glottis, 116.
Spastic spinal paralysis, 69.
Speech, centre for, 49; disturbances of, 49.
Sphygmography, 174.
Spinal cord, anterior horns, 40; posterior horns, 40.
Spinal muscular atrophy, 68.
Spinal puncture, 30.
Spine, curvature of, 122.
Spirals, Curschmann's, 149.
Spirilla, 264, 275.
Spirometry, 126.
Spleen, the, 106; dulness of, 106; enlargements of, in malaria, 106; enlargement of, in pneumonia,

106; enlargement of, in typhoid fever, 106; floating or movable, 106.
Splenic leucæmia, 251.
Spores, formation of, 264.
Sporozoa, 253.
Sputum, bacteria in, 146; black, 145; bloody, 142; blue, 145; cholesterin crystals in, 149; color of, 144; cylindrical epithelium in, 145; echinococcus in, 150; eosinophile cells in, 145; examination of, 141; fibrinous, 143; green, 144; heart-disease-cells in, 146; leptothrix in, 146; leucocytes in, 145; muco-purulent, 142; mucous, 142; ochre-yellow, 144; odor of, 144; pavement epithelium in, 145; purulent, 142; quantity of, 145; red, 145; red blood-cells in, 146; rubiginous, 144; rusty, 144; sarcina pulmonum in, 146; serous, 142; staining reactions of, 145; tyrosin crystals in, 150; yellow, 145.
Stadium decrementi, 18; incrementi, 18.
Stage of eruption, 19; incubation, 19.
Staphylococci, 267.
Status præsens, 2.
Stenosis of aorta, 178; bronchi, 111; larynx, 111; œsophagus, 72; trachea, 111.
Sternal line, 128.
Sterno-cleido-mastoid muscle, cramp of, 51.
Sterno-vertebral diameter, 124.
Sternum, 123; anomalies of, 123.
Stertor, 14.
Stomach, acid determination of, 86; ballooning of, 84; carcinoma of, 90; catarrh of, 90; contents, 85; contents, examination of, 85; dilatation of, 32, 84, 90; diseases of, 75; distention of, 84; free acid in, 86; inspection of, 81; motor power of, 89; pain in, 76; palpation of, 81; peptone in, 87; percussion of, 82; pressure and fulness in, 76; sarcina in, 76; total acidity of, 87; tumor of, 82; ulcer of, 90.

Stomatitis, 72.
Stool, black color of, 92; particles of tissue in, 93; green color of, 92; grayish-white color of, 92; mucus in, 92; pus in, 92.
Strabismus, convergent, 47; divergent, 47.
Strength, estimation of, 4.
Streptococci, 267.
Stridor, 110.
Stupor, 35.
Subfebrile temperature, 17.
Succussio Hippocratis, 140.
Suffusions, 12.
Sugar, qualitative test for, 193; quantitative test for, 196.
Sugillations, 12.
Sulphanilic acid, 201.
Sulphuric ether, 203.
Suppuration, fever of, 31.
Swallowing, sound of, 74.
Sweat, 12.
Symptoms, striking, 14.
Syphilis, cerebral, 67; laryngeal, 115; of lung, 159.
Systolic murmurs, 169; retraction of apex, 164; vesicular respiration, 136.

Tabes dorsalis, 60.
Tachycardia, 160, 171.
Tactile sense, 58; test of, 58.
Tænia cucumerina, 257; echinococcus, 257; mediocanellata, 255; nana, 257; solium, 254.
Tape-worms, 254.
Teeth, the, 71.
Teichmann's blood crystals, 249.
Temperature, hyperpyretic, 17; measurement of, 17; sense, 59; of skin, 12.
Tendon reflex, 53.
Test meals, 86.
Tetanus, 51; bacillus of, 277.
Tetany, 51.
Thermometers, 16.
Thoma-Zeiss blood-counting apparatus, 244.
Thomsen's disease, 52.

Thorax, auscultation of, 136; dilatation of, 123; flattening of, 123; inspection of, 122; measurements of, 124; narrowing of, 123; percussion of, 129.
Thread worms, 259.
Thrills, cardiac, 164.
Throat, the, 109.
Thrush, 72, 263.
Thyreoid gland, diseases of, 240.
Tic convulsif, 142.
Tidal air, 127.
Time of closure, 163.
Titration, 197.
Tongue, the, 71; inflammation of, 71.
Tonic convulsions, 50.
Tonsils, the, 109.
Topography of lung, 129.
Tracheal respiration, 111; stenosis, 111.
Transitory glycosuria, 195.
Trematoda, 258.
Tremors, 52.
Tremor, intention, 52; of eye, 53; senilis, 52.
Triacid mixture, 144.
Trichinæ spiralis, 260; intestines, 260; in muscle, 260.
Trichocephalus dispar, 260.
Trichophyton tonsurans, 262.
Tricuspid insufficiency, 178.
Trigeminus paralysis, 47.
Triple phosphates, 212.
Trismus, 51.
Trommer's test, 193.
Trousseau's phenomenon, 51.
Tubercle bacilli, 274; staining of, 266.
Tubercular meningitis, 67.
Tuberculin, 156.
Tuberculosis, acute miliary, 30.
Tumor of kidney, 225; of stomach, 82.
Tympanitic percussion note, 133.
Typhoid fever, 24; bacillus of, 270; uncertain stage, 24.
Typhus abdominis, 24; exanthematicus, 25; fever, 25.

Uffelmann's reagent, 86.
Ulcer of stomach, 90.
Ulcus ventriculi, 90.
Urates, acid sodium, 210.
Uræmia, 51.
Urea, 204; demonstration of, 205; quantitative determination of, 205 seq.
Uric acid, 207; calculi, 226; crystals, 210.
Urine, albumin in, 185; albumoses in, 188; acid, 183; acid, sediment in, 210; alkaline reaction of, 184; alkaline sediment in, 212; alloxan bodies in, 208; ammonia in, 202; antifibrin in, 218; antipyrin in, 218; arsenic in, 217; biliary coloring matter in, 191; bilirubin in, 191; blood in, 189; bromine in, 216; carbolic acid in, 209, 218; carbonates in, 202; casts in, 214; chlorides in, 202; cloudiness of, 183; color of, 182; cystin in, 211; diminution in amount of, 181; examination of, 181; fat in, 202; hæmoglobin in, 189, 191; hippuric acid in, 208; hydrobilirubin in, 192; in fever, 18; in heart-disease, 176; increased amount of, 182; indican in, 191, 208; test for indican in, 209; iodine in, 216; iron in, 217; kreatinin in, 208; lead in, 218; leucin in, 212; leucocytes in, 213; melanin in, 202; mercury in, 218; micro-organisms in, 216; naphthalin in, 219; organised sediment in, 213; oxalic acid in, 208; peptone in, 188; phenacetine in, 218; phenol in, 209; phosphates in, 202, 212; potassium in, 204; pus in, 213, 216; quantity of, 181; quinine in, 219; reaction of, 183; red blood-cells in, 213; renal epithelium in, 214; rhubarb in, 219; salicylic acid in, 218; santonin in, 219; sediment in, 210; senna in, 219; sodium in, 204; specific gravity of, 182; sugar in, 192; sulphates in, 202; sulphuretted hydrogen in, 202; tannin in, 219;

turpentine in, 219; tyrosin crystals in, 212; unorganized sediment in, 210; urobilin in, 192.
Urobilin, test for, 192.

VALVULAR cardiac lesions, 177.
Van Deen's test for blood, 189.
Varicella, 27.
Variola, 26.
Varioloid, 27.
Venous pulse, 164.
Ventricular voice, 109.
Vermes, 253.
Vertebra prominens, 127.
Vertigo, 37.
Vesicular respiration, 136.
Vessels, auscultation of, 170.
Vibriones, 270.
Vital capacity, 126.
Vocal cords, cadaveric position of, 119; paralysis of, 116 seq.
Vocal fremitus, 141.
Voice, 109; auscultation of, 141; threefold splitting of, 110.
Volumen pulmonum auctum, 152.
Vomiting, 77; fæcal, 80.
Vomitus matutinus, 80; microscopic examination of, 79.

WANDERING kidney, 225.
Waxy casts, 214.
White blood-cells, 244; number of, 244.
Whooping cough, 122.
Widal's reaction, 272.
Wintrich's change of sound, 135.
Worms, 253.
Wound fever, 31.

XANTHIN bodies, 208; stones, 227.

YELLOW sputum, 145.

ZOOGLÆA, 264.

A
List of Books on Medicine, Etc.

ALLBUTT. — **A System of Medicine.** By many Writers. Edited by THOMAS CLIFFORD ALLBUTT, M.A., M.D., LL.D., F.R.C.P., F.R.S., F.L.S., F.S.A., Regius Professor of Physics in the University of Cambridge, etc. In six volumes, medium 8vo. To be issued quarterly, beginning July, 1896. (*Orders received for sets only.*)

Vol. I. **Prolegomena and Infectious Diseases.** Cloth, $5.00. Half Leather, $6.00. Illustrated with 33 Figures in the Text, 13 Charts of Death Rate, Temperature, etc., and a Colored Plate.

Vol. II. **Infective Diseases.** Cloth, $5.00. Half Leather, $6.00. With 77 Figures, 6 Charts, Map and Colored Plate.

Vol. III. **Diseases of Obscure Causation; Diseases of Alimentation and Excretion.** Cloth, $5.00. Half Leather, $6.00.

Vol. IV is nearly ready; Vols. V and VI in preparation.

"This is the beginning of an elaborate 'system' which is destined to become a very important addition to our literature. The work is a pioneer in many directions." — *The Journal of the American Medical Association.*

"Although it must of necessity follow that a work which has such a list of contributors as this has, will be of a superior order, yet in anticipation one rather underestimates than overestimates its worth. As he reads, however, he has ample evidence of the superiority of the book and perceives how unique in many respects it is. To judge from this, the first volume, it is not too much to say that it, more than most works of similar compass, deserves the title of system, for it is comprehensive, scientific, and systematic in the highest degree. . . . The introduction, in particular, merits the warmest terms of admiration. It is written by the editor himself, and for greatness of thought, broad comprehensiveness and beauty of language, is a most able production." — *The Medical Journal,* New York.

ALLBUTT and PLAYFAIR. — **A System of Gynæcology.** By many Authors. Edited by T. C. ALLBUTT, Editor of *A System of Medicine,* and L. S. PLAYFAIR, Professor of Gynæcology, King's College. Cloth, $6.00. Half Leather, $7.00.

This is uniform with the six-volume *System of Medicine,* under the general editorship of Dr. ALLBUTT. To subscribers of that *System,* entire (7 volumes including the Gynæcology), the price is Cloth, $5.00. Half Leather, $6.00.

BALFOUR. — **The Senile Heart.** Its Symptoms, Sequelæ, and Treatment. By G. W. BALFOUR, M.D. 12mo. Cloth, $1.50.

"A very clearly expressed and readable treatise upon a subject of great interest and importance. We can heartily commend this work." — *Medical Record*, New York.

BARR. — **Manual of Diseases of the Ear, including those of the Nose and Throat in Relation to the Ear.** For the use of Students and Practitioners of Medicine. By THOMAS BARR, M.D., Lecturer on Diseases of the Ear, Glasgow University. Second Edition, largely rewritten. With 229 Illustrations. $3.50.

BRUNTON. — **On Disorders of Digestion.** Their Consequences and Treatment. By T. LAUDER BRUNTON, F.R.S. $2.50.

—— **Lectures on the Action of Medicine.** Lectures on Pharmacology and Therapeutics, delivered at St. Bartholomew's Hospital, 1896. By T. LAUDER BRUNTON, F.R.S. 8vo. Cloth, $4.00.

CLELAND and McKAY. — **Anatomy of the Human Body.** By Dr. JOHN CLELAND, Professor of Anatomy in the University of Glasgow, and Dr. JOHN YULE MCKAY, Professor of Anatomy in University College, Dundee. 8vo. Cloth, $6.50.

The object of the authors has been to produce a work that should be accurate, comprehensive, up to date, and yet sufficiently brief for the use of students.

FOSTER. — **A Text-book of Physiology.** By MICHAEL FOSTER, M.A., M.D., LL.D., F.R.S., Professor of Physiology, University of Cambridge, Fellow of Trinity College, Cambridge. 8vo. Illustrated. Sixth edition, largely revised.

   Part I. Blood; the Tissues of Movement; Vascular Mechanism. $2.60.

   Part II. The Tissues of Chemical Action; Nutrition. *New Edition in Press.* $2.60.

   Part III. The Central Nervous System. *New Edition.* $2.50.

   Part IV. The Central Nervous System (*concluded*); the Tissues and Mechanisms of Reproduction. $2.00.

   Part V. (Appendix.) The Chemical Basis of the Animal Body. By A. SHERIDAN LEA, ScD., F.R.S. $1.75.

   Text-book of Physiology. Revised and Abridged from the work described above to one volume. 8vo. Cloth, $5.00. Sheep, $6.00.

"We have used this valuable work (for the most part in the five-volume edition) since its first publication, and will continue to do so." — V. C. VAUGHAN, *Dean of Medical Dept., University of Michigan.*

"The abridgment is much better adapted to the use of medical students than the five-volume edition, and we are recommending it to our students as the best existing English text-book of physiology for their use."— FREDERIC S. LEE, *Adjunct Professor of Physiology, Columbia University.*

"I am much pleased with the form in which this edition is issued, since it makes it more suitable as a text-book for students, and removes some of the objections which have hitherto attended the use of the former editions. It is compact so far as its subject-matter goes, and the abridgment is a great improvement. I shall take pleasure in pointing out this fact to my students."— CHARLES D. SMITH, *Professor of Physiology, Bowdoin College.*

"I have for several years recommended this text-book to my classes in the Medical Department of the University of California, and have always regretted that the edition they were compelled to purchase was not an authorized one. I shall take pleasure hereafter in recommending your edition to my students."— A. A. D'ANCONA, *Professor of Physiology, University of California.*

FOTHERGILL.— **The Practitioner's Handbook of Treatment**; or, The Principles of Therapeutics. By the late J. MILNER FOTHERGILL. Edited, and in great part rewritten by WM. MURRELL. *Fourth Edition.* 8vo. Cloth, $5.00.

FOTHERGILL.— **Manual of Midwifery.** For the Use of Students and Practitioners. By W. E. FOTHERGILL, M.A., B.Sc., M.B., C.M.; Buchanan Scholar in Midwifery, University of Edinburgh, etc. With Double Colored Plate and Sixty-nine Illustrations in the text. 12mo. Cloth, $2.25.

"This work is a modern handbook on obstetrics, and not only the student, but the practitioner and some writers on allied topics may gain a large quota of knowledge from its reading. The book is a safe teaching guide and a most excellent handbook."— *The Medical Journal,* New York.

FROST.— **The Fundus Oculi.** With an Ophthalmoscopic Atlas illustrating its Physiological and Pathological Conditions. By W. ADAMS FROST, F.R.C.S., Ophthalmic Surgeon, St. George's Hospital, etc. 4to. Cloth, gilt top, $18.00.

"The direct, concise, and lucid manner in which the descriptions of the various conditions are given is truly admirable. Exhaustive without being verbose, complete in facts without being confusing, the conception and completion of the argument leaves little to be desired. Too much cannot be said in praise of the colored plates."— *The Medical Record,* New York.

HAMILTON.— **A Systematic and Practical Text-book of Pathology.** By D. J. HAMILTON, M.D. Cloth. 8vo.

Vol. I. Technical. **General Pathological Processes. Diseases of Special Organs.** $6.25.

Vol. II. **Diseases of Special Organs** (*continued*). **Bacteriology,** etc. 2 parts. Each, $5.00.

"This is beyond question the most complete work on pathology in the English language to day. The author has accomplished his laborious task most successfully. We cannot better criticise it than by saying it is beyond criticism."— *Canada Medical Record.*

HAWKINS. — **On Diseases of the Vermiform Appendix.** With a Consideration of the Symptoms and Treatment of the Resulting Forms of Peritonitis. By H. P. HAWKINS, F.R.C.P. 8vo. $2.25.

"An excellent review of the subject, particularly in its pathological, statistical, and therapeutical bearings." — *The Medical Record*, New York.

ILLOWAY. — **Constipation in Adults and Children.** With special reference to Habitual Constipation and its most successful Treatment by the Mechanical Methods. By H. ILLOWAY, M.D., formerly Professor of the Diseases of Children, Cincinnati College of Medicine and Surgery. With many plates and illustrations. 8vo. Cloth, $4.00.

"We are extremely glad to note the appearance of a work of such great practical importance as this produced by Dr. Illoway. . . . Indeed, the author's description of the mechanical methods for overcoming habitual constipation and their application deserves general reading." — *Memphis Medical Monthly*

"The author of this book has supplied a much needed work on the subject of constipation. The frequency and unpleasant complications of this pathological condition make a demand for a reliable and instructive treatise thereon: such is found in this book." — *Eclectic Medical Journal.*

"It is a very valuable contribution, and, in fact, the only complete monograph on the subject in recent days of which we have knowledge." — *Virginia Medical Monthly.*

JENNER. — **Lectures and Essays on Fevers and Diphtheria, 1849-1879.** By SIR WILLIAM JENNER. 8vo. $4.00.

"This volume was a fitting exemplar of the careful and scientific work that has placed the author in the foremost rank of his profession. It cannot fail to prove interesting to physicians." — *The Medical Journal*, New York.

—— **Clinical Lectures and Essays on Rickets, Tuberculosis, Abdominal Tumors, and Other Subjects.** 8vo. $4.00.

"Rarely are collected in a single volume of small size so much of valuable clinical material as can be found herein. The ripe experience, broad views, and soundness of judgment of the author are manifest in a striking degree, and make the subject not only interesting but instructive. The commendable endeavor to make all the varied pathological conditions present in these common and wide-spread diseases accord with the well-recognized symptoms so often presented, rounds out the subjects with a completeness of description and practical interest which cannot fail to give new points of view to every reader and new ideas to every student." — *The Medical Record*, New York.

KIMBER. — **Text-book of Anatomy and Physiology for Nurses.** Compiled by DIANA CLIFFORD KIMBER, graduate of Bellevue Training School; Assistant Superintendent New York City Training School, Blackwell's Island, N. Y.; formerly Assistant Superintendent Illinois Training School, Chicago. Fully Illustrated. 8vo. $2.50.

"From her long experience in teaching classes the author knows exactly what nurses need and how much can be reasonably given them in the short space of two years' time, and for the assistance of the inexperienced teacher her book is arranged in lessons covering the first or junior year. The subjects are presented with sincerity and distinction, and illustrated by cuts and plates of unusual merit." *The Trained Nurse.*

"A happy mingling of theoretical knowledge with the requisite technical instruction required by practical work seems to have been attained in this volume. The lessons are progressive and so arranged as to unfold anatomy and physiology to nurses in a natural, helpful manner. The physiological discussions follow easily upon anatomical descriptions in such a manner as to make an impression upon the student.

"The volume is beautifully printed and well illustrated, and the author deserves much credit for her work as compiler. The glossary and index are especially satisfactory." — *The Johns Hopkins Hospital Bulletin.*

KOCHER. — **Text-book of Operative Surgery.** By Dr. THEODOR KOCHER, Professor of Surgery and Director of the Surgical Clinic in the University of Bern. Translated with the Special Authority of the Author from the Second Revised and Enlarged German Edition by HAROLD J. STILES, M.B., F.R.C.S. Edin., Senior Demonstrator of Surgery, University of Edinburgh, etc. With 185 Illustrations. 8vo. $3.50. *New edition in preparation.*

"This is a translation of the second edition of Dr. Kocher's excellent treatise on operative surgery, which has already won for itself recognition as the standard work on this subject. The translation has been well done, and the illustrations are admirable." — *The Medical Record,* New York.

MACDONALD. — **A Treatise on Diseases of the Nose and its Accessory Cavities.** By GREVILLE MACDONALD, M.D. Second Edition. 8vo. $2.50.

"The author has adopted a method of indexing his work that is sure to commend it to general favor — table of contents, chapter-indexing, and marginal notes. The busy man is always grateful when he is enabled by any such device to find at a glance the division of the subject that particularly interests him, and when, moreover, he finds, on consulting the text, that the author has written, so to speak, with the patient before him, his attention and confidence are entirely gained and he follows with the closest attention the lessons of one who has seen and understood. Macdonald's style is admirably adapted to his subject, which latter he keeps, without deviation, constantly before his eyes. We cannot too highly recommend the treatise." — *The Medical Journal,* New York.

MACEWEN. — **Pyogenic Infective Diseases of the Brain and the Spinal Cord, Meningitis, Abscess of Brain, Infective Sinus Thrombosis.** By WILLIAM MACEWEN, M.D. (Glasgow). 8vo. Buckram, $6.00.

"The careful, precise methods followed by the author, his thorough familiarity with cerebral anatomy and surgery, and his habit of waiting for time to demonstrate the value of what he has done, make this volume the most valuable contribution to the surgery of the brain that has appeared in several years. The illustrations are magnificent, on a par with those in his beautiful *Atlas of Head Sections.* The volume must be of great interest to the neurologist, the aurist, and the surgeon." — *The Medical Journal,* New York.

—— **Atlas of Head Sections.** 53 Engraved Copperplates of Frozen Sections of the Head and 53 Key Plates with Descriptive Text. By WILLIAM MACEWEN, M.D. Folio. Bound in Buckram, $21.00.

"The *Atlas* should certainly be in the hands of every surgeon who aspires to enter the field of brain-surgery, as a careful inspection of these plates will teach more than many volumes written upon the subject." — *The Medical Record,* New York.

MINOT. — **Human Embryology.** By CHARLES SEDGWICK MINOT, Professor of Histology and Human Embryology, Harvard Medical School, Boston. With 463 Illustrations. 8vo. $6.00.

PLAYFAIR. — **A System of Gynæcology**, under the editorship of Dr. W. S. PLAYFAIR, is issued also as an independent volume, and can be purchased separately by those who do not care for the *System of Medicine*, edited by Dr. ALLBUTT, in connection with which it is issued. If bought apart from the *System of Medicine* the price will be, Cloth, $6.00, Half Leather, $7.00.

STARR. — **Atlas of Nerve Cells.** By M. ALLEN STARR, M.D., Ph.D., Professor of Diseases of the Mind and Nervous System, College of Physicians and Surgeons, Medical Department, Columbia College; Consulting Neurologist to the Presbyterian and Orthopædic Hospitals, and to the New York Eye and Ear Infirmary. With the coöperation of OLIVER S. STRONG, A.M., Ph.D., Tutor in Biology, Columbia College, and EDWARD LEAMING. Illustrated with 53 Albert-type Plates and 13 Diagrams. *Columbia University Press.* Royal 4to. Cloth, $10.00.

"The paper, typography, and beautifully reproduced plates of this quarto atlas give it the appearance of an *edition de luxe*. . . . The explanation of the plates and the exposition of the subject display the clear, concise, and comprehensive style characteristic of Dr. Starr." — *The Medical Record*, New York.

"Dr. Starr's work will enable the numerous students and practitioners who cannot undertake original research to understand completely the new facts which have been disclosed. The reproductions of micro-photographs leave nothing to be desired, and the details of nerve structure are shown with the minutest accuracy and with the most perfect clearness. . . . In the opinion of the London *Times*, 'their work will be indispensable to all who desire to become familiar with the anatomy of the nervous system.'" — *The Evening Sun*, New York.

STEPHENSON. — **Epidemic Ophthalmia.** Its Symptoms, Diagnosis and Management. With papers upon Allied Subjects. By SIDNEY STEPHENSON, M.B., F.R.C.S. Ed. 8vo. Cloth, $3.00.

"This is an elaborate, scientific and practical treatise on a subject of wide economic and social interest. It is divided into four sections. The first treats of epidemic ophthalmia, its symptoms, diagnosis, and management. The second consists mainly of a clinical inquiry into the prevalence and significance of the follicular granulation of the conjunctiva. The third is devoted to the treatment of trachoma and its complications, and the fourth to the treatment of follicular conjunctivitis. There is a good index of subjects and authors, also a list of the works consulted." — *The Medical Journal*, New York.

THOMA. — **Text-book of General Pathology and Pathological Anatomy.** By RICHARD THOMA. Translated by ALEXANDER BRUCE, M.D., F.R.C.P.E., etc., Lecturer on Pathology, Surgeon's Hall, Edinburgh, etc. With 436 Illustrations. Vol. I. Imp. 8vo. Cloth, $7.00.

"The first volume of this well-known work constitutes one of the most valuable contributions to the subject in any language, and English-speaking readers are to be congratulated upon its translation from the German. . . . Too much praise can hardly be given to the work of the translator. The diction is always pure and readily intelligible, the author's meaning reappears accurately, and there is an entire absence of the traces of foreign style so frequently found in English translations of German medical works. Finally, the publishers have presented a volume which in binding, paper, printing, and general detail is of a very high order." — *The Medical Journal*, New York.

TUBBY. — **Deformities.** A Treatise on Orthopædic Surgery, for Practitioners and Advanced Students. By A. H. TUBBY, M.S., F.R.C.S. Eng., Assistant-Surgeon to and in charge of the Orthopædic Department, Westminster Hospital. Illustrated with 15 Plates and 302 Figures, of which 200 are original, and by Notes of 100 Cases. Large 8vo. Cloth, $5.50.

"This volume is the outcome of several years' work, by the author, at the National Orthopædic Hospital, the Evelyn Hospital for Sick Children, and for some time in the Orthopædic Department at the Westminster Hospital. The author, however, has not only made a record of his own work, but has given a fair account of the deformities as at present understood. It is pleasing to note that he has quoted freely from Bradford and Lovett of this country, and pays a graceful tribute to our Orthopædic Association by saying: 'Above all, I cannot omit to express my sense of indebtedness to the many admirable writers who have recorded their experiences in the Transactions of the American Orthopædic Association.' . . .

"The book is timely and, although conservative, is fully up to date. . . . We commend the book as one being in every way satisfactory." — *The Journal of the American Medical Association*.

UNNA. — **The Histopathology of the Diseases of the Skin.** By Dr. P. G. UNNA. Translated from the German, aided by the Author, by NORMAN WALKER, M.D., F.R.C.P., Assist. Physician in Dermatology to the Royal Infirmary, Edinburgh. With a Double Colored Plate containing Nineteen Illustrations and Forty-two Additional Illustrations in the Text. Cloth, $10.50.

"Dr. Unna has given us a great book; in its twelve hundred pages there is much that is valuable and there is considerable that is new. . . . It is sufficient to say that the book will no doubt prove a valuable addition to the shelves of those who are working in the line of its subject-matter." — *The Medical Journal*, New York.

WARING. — **Diseases of the Liver, Gall Bladder, and Biliary System.** Their Pathology, Diagnosis, and Surgical Treatment. By H. J. WARING, M.S., B.Sc. Lond., F.R.C.S., St. Bartholomew's Hospital, London. 8vo. $3.75.

WATSON. — **Practical Handbook of the Diseases of the Eye.** Containing Nine Colored Plates and Twenty-four Illustrations in the Text. By D. CHALMERS WATSON, M.B., C.M., Ophthalmic Surgeon, Marshall Street Dispensary, Edinburgh. 16mo. $1.60.

WEBSTER. — **Ectopic Pregnancy.** Its Etiology, Classification, Embryology, Diagnosis, and Treatment. By J. CLARENCE WEBSTER, M.D., F.R.C.P. Ed., Assistant of the Professor of Midwifery, etc., University of Edinburgh. With Eighty Illustrations. 8vo. Cloth, $3.75.

WILSON. — **An Atlas of the Fertilization and Karyokinesis of the Ovum.** By EDMUND B. WILSON, Ph.D., Professor in Invertebrate Zoölogy in Columbia College, with the Coöperation of EDWARD LEAMING, M.D., F.R.P.S., Instructor in Photography at the College of Physicians and Surgeons, Columbia College. 4to. $4.00.

"This work is of a very high order, and both by its merit and its opportuneness is a noteworthy contribution to science. . . . The work takes its place at once as a classic, and is certainly one of the most notable productions of pure science which have appeared in America. It will be valuable to every biologist, be he botanist or zoölogist, be he investigator or teacher." — *Science.*

The Cell in Development and Inheritance. By EDMUND B. WILSON, Ph.D. With Illustrations. $3.00.

ZIEGLER. — **A Text-book of Pathological Anatomy and Pathogenesis.** With Illustrations. By ERNST ZIEGLER, Professor of Pathology in the University of Freiburg. Translated by DONALD MACALLISTER, M.A., M.D., Cambridge, and HENRY W. CATTELL, A.M., M.D., Demonstrator of Morbid Anatomy, University of Pennsylvania. 8vo.

Vol. I. **General Pathological Anatomy.** New Edition. *In Preparation.*

Vol. II. **Special Pathological Anatomy.** New Edition. Thoroughly Revised. Sections I-VIII. $4.00.

This is a thoroughly revised and entirely reset edition of the standard text-book. In its revision the latest German edition has been consulted throughout and the book is practically a new translation. A valuable index, etc., has been added by DR. CATTELL.

THE SAME. Sections IX-XV. *Just Ready.* $4.00.

www.ingramcontent.com/pod-product-compliance
Lightning Source LLC
Chambersburg PA
CBHW030805230426
43667CB00008B/1074